ENVIRONMENTAL HAZARDS

MARINE POLLUTION

Man has lost the capacity to foresee and forestall.
He will end by destroying the earth.
—Albert Schweitzer, as quoted by Rachel Carson in the dedication of *Silent Spring*

It is the same old story—jobs before safety, profits before protection of the environment, a short-sighted policy that, in the name of development . . . is prepared to sacrifice the future for immediate material gain.
—K. A. Gourlay, in *Poisoners of the Seas*

The Earth is one but the world is not. We all depend on one biosphere for sustaining our lives. Yet each community, each country, strives for survival and prosperity with little regard for its impact on others.
—Norwegian Prime Minister Gro Harlem Brundtland, in *Our Common Future*

The Oceans gave us life. It's time we returned the favor.
—American Oceans Campaign

ENVIRONMENTAL HAZARDS

MARINE POLLUTION

Martha Gorman

CONTEMPORARY WORLD ISSUES

Santa Barbara, California
Denver, Colorado
Oxford, England

Library of Congress Cataloging-in-Publication Data

Gorman, Martha, 1952–
Marine pollution : a reference handbook / Martha Gorman.
p. cm.—(Contemporary world issues)
Includes bibliographical references and index.
1. Marine pollution. I. Title. II. Series.
GC1085.G67 1993 363.739409162—dc20 93-39789
ISBN 0-87436-641-0 (alk. paper)

00 99 98 97 96 95 94 10 9 8 7 6 5 4 3 2 (hc)

ABC-CLIO, Inc.
130 Cremona Drive, P.O. Box 1911
Santa Barbara, California 93116-1911

This book is printed on acid-free paper ♾.
Manufactured in the United States of America.

Contents

Preface, ix

Acknowledgments, xiii

1 Introduction, 1

Earth—The Water Planet, 2
What Is Marine Pollution?, 3
Sewage, 5
Treating Sewage, 5
How Sewage Pollutes, 6
Monitoring Sewage, 6
Cleaning Up Sewage Pollution, 7
Marine Debris, 7
Types of Marine Debris, 7
How Marine Debris Enters the Ocean, 8
The Effects of Marine Debris on Living Organisms, 9
Monitoring Marine Debris, 9
Cleaning Up Marine Debris, 10
Solutions, 10
Toxic Chemicals, 11
How Toxic Chemicals Enter the Ocean, 12
The Effects of Toxic Chemicals on Living Organisms, 13
Cleaning Up Toxic Chemicals, 14
Solutions, 14
Heavy Metals, 15
How Heavy Metals Enter the Ocean, 15
The Effects of Heavy Metals on Living Organisms, 16

Monitoring Heavy Metal Pollution, 17
Cleaning Up Heavy Metals, 17
Solutions, 18

Oil, 18

How Oil Enters the Ocean, 19
The Effects of Oil Pollution on Living Organisms, 20
Monitoring Oil Pollution, 20
Cleaning Up Oil Pollution, 21
Solutions, 22

Radioactive Materials, 23

How Radioactive Materials Enter the Ocean, 24
Disposing of Radioactive Wastes, 25
The Effects of Radiation on Living Organisms, 25
Monitoring Radiation, 26
Cleaning Up Radioactive Wastes, 27
Solutions, 27

Our Changing Perceptions of Marine Pollution, 28

Conclusions, 30

2 Chronology, 31

3 Biographical Sketches, 53

4 Legal Framework, Facts and Data, and Points of View, 67

Legal Framework, 68

International Conventions and Agreements, 68
U.S. Marine Pollution Laws, 92

Facts, Statistics, Tables, and Figures, 106

General Facts and Statistics, 106
Sewage, 108
Marine Debris, 113
Toxic Chemicals, 114
Heavy Metals, 119
Oil, 119
Radioactive Materials, 122

Points of View, 129

What Is Marine Pollution?, 129
Sewage, 132
Marine Debris, 134
Toxic Chemicals, 135
Heavy Metals, 137
Oil, 138
Radioactive Materials, 140
Our Changing Perceptions of Marine Pollution, 141

5 Directory of Organizations, 149

General Organizations, 149

Libraries, 183

6 Selected Print Resources, 185

Reference Materials, 186

Journals, 190

Government Publications, 191

Books and Articles, 195

General, 195
Sewage, 204
Marine Debris, 205
Toxic Chemicals, 206
Heavy Metals, 207
Oil, 207
Radioactive Materials, 209

7 Selected Nonprint Resources, 211

Audio and Visual, 211

Computer Networks and Databases, 227

Television and Radio Series, 230

Glossary, 233

Index, 241

Preface

As a child, every spring I drove south with my family to St. Petersburg, Florida to visit my grandparents. They lived just a block from Tampa Bay, and I spent many enjoyable hours diving off a small wooden dock that jutted several yards "out to sea." Although I lived just miles from Lake Erie and the Niagara River, only rarely were we allowed to swim on the American side and finally even the Canadian beaches began to close. But that was just a lake, I remember thinking. Lakes can get dirty. But the ocean in its immensity was surely not vulnerable to such damage. By the time I was 11, however, a sign was posted forbidding me to swim . . . "Even the ocean can die," I remember thinking.

In the course of researching this book, I discovered that there is no single source of information on marine pollution. Data are dispersed under topics such as "water" and "oceans." One must search under "sewage" and "oil" and "PCBs." Important sources are hidden under the keywords "environmentalism" and "conservation." It seems as though either all kinds of water are thrown in together, or all types of pollution. "Marine pollution" as a subject heading hardly exists yet.

This should not be surprising. Marine pollution is an extremely recent concept—less than half a century old, really. Marine oil pollution was recognized as an environmental problem in Europe during the early 1950s, and sewage disposal methodologies came under scrutiny in the U.S. at about the same time. The American public began to question the wisdom of ocean dumping in 1959 when a drum containing radioactive waste washed ashore in California. This incident may mark the start of the environmental movement in the U.S., although the focus of that movement was on land-based issues. Oil pollution remained the foremost concern of those working in the area newly termed "marine pollution" during the 1960s, the decade during which both the National Oceanographic and Atmospheric Administration (NOAA) and the Sea Grant Program were founded, PCBs were identified and discovered in penguin eggs, the *Torrey Canyon* spill occurred, and the President's Panel on Oil Spills met for the first time.

Toxic chemical and heavy metal pollution concerns grew during the 1970s, when both dumping and regulations against dumping proliferated. During the 1980s, marine debris and the disposal of radioactive waste from nuclear power plants were added to the growing list of marine pollution issues. The regulations enacted during the 1970s in the U.S. were undermined during 1980s. The 1990s, however, are already exhibiting a greater environmental awareness than that seen since the 1960s. It can only be hoped that further environmental catastrophes like the *Exxon Valdez* spill will not be needed to galvanize a global resolve to preserve the oceans.

Although the pollution of groundwater and inland fresh water certainly contributes to marine pollution, in the interest of space, this book deals only with the pollution of the planet's salt waters and estuaries. And although the book focuses on the waters surrounding the United States, this does not mean the problem is limited to this geographic area, or even that these waters are in the worst shape. Rather, the sheer immensity of information concerning this area, space limitations, and the probable needs of readers demanded this focus. (However, numerous international agencies are listed in Chapter 5, as are several key European organizations.) Adventurous readers interested in discovering that the North Sea and the Baltic compete with the New York Bight as the most polluted marine environments in the world are encouraged to communicate with these extremely helpful institutions.

This book is organized to provide ready access to a broad range of information about marine pollution. Chapter 1 describes the ocean's place in the biosphere and defines the scope of marine pollution. Then six major categories of marine pollutants are dealt with in depth: sewage, marine debris, toxic chemicals, heavy metals, oil, and radioactive materials. Chapter 2 provides a chronology of events, starting in the fourth century but focusing on the last half of the twentieth century. Chapter 3 provides brief biographies of twenty key individuals in the field of marine studies and pollution, ranging from Jacques-Yves Cousteau to a corporate whistleblower who lost his job reporting serious ocean dumping violations. Chapter 4 provides the legal framework for marine pollution prevention, presents tables, figures, and facts that graphically illustrate key aspects of the six categories of marine pollution, and puts the issues into perspective by excerpting and quoting from key documents and speeches. Chapter 5 lists organizations and associations around the world, ranging from government laboratories to activist groups. Chapter 6 gives print resources, and Chapter 7 lists nonprint resources. Finally, a glossary provides definitions for some of the technical terms used.

Our recent understanding of the ocean as a single body of water and of our planet as a single, delicately balanced ecosystem has radically changed our perception of marine pollution. Now it is time to act on this new knowledge—before it is too late.

Acknowledgments

Although many individuals contributed greatly to the compilation of this book, I would like to especially thank my editor, Henry Rasof, for getting me on track and keeping me there. Without his assistance and persistence, this book would not have been completed. Martha Whitt, also of ABC-CLIO, was generous in her support during the final production stage.

I would also like to salute my husband, Stuart McKeen, my son, Skylar, and my daughter, Starin, for cheerfully putting up with the long hours and poor temper that this two-year project entailed. I hope they consider the sacrifices—including the deletion of seafood from our family diet—worthwhile.

Finally, I would like to give thanks to my good friend, Beverly Vanderkooy, of the State University of New York at Buffalo Lockwood Memorial Library, for her unstinting assistance during the initial research phase of the project; to the staff at the Pell Library and the Sea Grant Depository at the University of Rhode Island; to Brian and Mary Heikes; and to the reference staff of the Boulder (Colorado) Public Library, for their encouragement and assistance.

Should this book spark the interest of but one individual to dedicate time and energy to protect the sea and the planet of which it is an integral part, the project shall have been worthwhile.

1

Introduction

ON THE SHORE OF THE SHIRANUI SEA, on the east coast of the island of Kyushu in Japan, lies the fishing and industrial town of Minamata. In the 1950s local fish was an extremely important part of the diet of Minamata's citizens, who then numbered 50,000. But the town's most profitable business and largest employer was not the fishing industry, but the Shin-Nihon Chisso Hiryo chemical factory.

In 1953, disturbing symptoms began to appear among the people of Minamata. Some experienced numbness in their lips; others had numbness in their arms and legs as well, and they walked as if drunk. They began to lose their peripheral vision, and some had seizures. Many died.

By 1956, Minamata Disease, as it was called, had become epidemic. Although at first doctors thought it was a contagious form of meningitis, they finally realized—after observing the village cats, who also suffered from the ailment—that the townspeople were suffering from heavy metal poisoning. But what metal, and how was it getting into their bodies?

Researchers discovered that since the 1930s the chemical factory had been generating mercury as a by-product of one of its manufacturing processes, and had disposed of it by dumping it into a river that empties into Minamata Bay. The victims of Minamata Disease consumed as much as 14 milligrams of mercury—an enormous amount—merely by eating the fish they caught in the bay at their doorstep. Fishing in Minamata Bay was banned in 1957; no additional people became ill that year.

Although the factory repeatedly denied any role in the deaths, in 1969 it was finally proven beyond doubt that mercury had caused the poisoning. Twenty-eight families sued for damages.

The Minamata tragedy is important because it demonstrated, for the first time, a clear-cut connection between marine industrial pollution and human health hazards. It also showed that using the ocean as an easy, inexpensive dumping ground for waste, as is done almost everywhere in the world, may have unforeseen and possibly disastrous consequences. In Minamata, the dependence of the townspeople on locally caught fish brought these consequences to light much sooner than might otherwise have been the case.

Earth—The Water Planet

Imagine you are traveling on the space shuttle. As you enter into orbit around the earth, you see a blue planet—a water planet. Contrary to what many textbooks once claimed, there are not "seven seas." From the shuttle, you see that all the gulfs and seas and oceans on the face of the earth make up a single, interconnected world ocean.

What you would *not* be able to see is the enormous variety of what lies below the ocean's shimmering blue surface. There is an entire world of fish and sharks, whales and dolphins, huge expanses of seaweed, and other sea life. There are underwater volcanoes, great mountain ranges, deep valleys, and other striking geological features. The volume of water in the ocean is so huge that if the ocean floor were uniformly flat around the globe, the ocean would be 11,600 feet deep—as deep as many mountains are high! But the ocean floor is anything but uniform. Some deep ocean valleys extend 30,000 feet below the watery surface, a greater distance below sea level than Mount Everest is above it.

The ocean is about 2.5 billion years old, or nearly half the age of the earth. It covers 71 percent of the earth's surface and contains 318 million cubic miles of water. Like the atmosphere, the ocean is a dynamic entity, always changing. Together, the atmosphere and the ocean form a single massive, interrelated system that affects nearly every aspect of life on the planet. Because the ocean and the atmosphere are so tightly coupled, any substance we humans put into the air almost inevitably ends up in the sea (often carried there as particles within raindrops). And what goes into the ocean often winds up in our food or, through ocean evaporation and rainfall, on the land.

Within this massive system, the ocean functions much like a global thermostat. Water has an enormous capacity to store heat; if there were no ocean, the earth's climate would be more like that of the moon—all extremes of heat and cold.

The ocean is also the foundation of the planet's food chain. Plankton, the microorganisms that inhabit the sea's fertile surface waters, provide food directly or indirectly for most of the planet's creatures, including of course the fish that are an important part of the human diet.

Furthermore, sea water is not *just* water. It has suspended in it an enormous number of chemical elements and compounds, in a highly stable balance. One of the most common of these chemical compounds, salt, provides a good example of this remarkable balance—the ocean's salinity has remained constant for at least 500 million years. Until very recently, in fact, the proportion of nearly all the elements in sea water—and sea water contains in dissolved form nearly every known element—was constant.

Thus, the ocean turns out to be much more complicated and surprising than our space shuttle view would have led us to believe. Our relatively recent understanding of the ocean as a single body of water and of the planet as a single, delicately balanced ecosystem has radically changed our perception of marine pollution over the past several decades.

What Is Marine Pollution?

Humans began polluting the Earth's seas the first time someone threw a half-gnawed bone or a banana peel into the waves, but the first large-scale impact of humanity on the balance of the oceans began with the rise of agriculture during the earliest civilizations. As permanent, settled communities replaced nomadic tribes, less human and animal waste was returned to the land and more was concentrated in the nearest body of water, either as runoff or as sewage.

Increased nutrient levels in the sea initially improved fishing. For instance, in the fourth century in the waters near

Constantinople, sewage apparently produced a plankton population explosion that attracted whales to the area. Ten centuries later, the population in The Netherlands and the growing metropolis of London also poured sufficient sewage into the sea to increase plankton, which in turn caused greatly increased herring catches. Since there were no obvious negative consequences, no one considered this to be "pollution."

The term "marine pollution" was not widely used until 1967, when the tanker *Torrey Canyon* spilled more than 36 million gallons of crude oil just 20 miles off the coast of Cornwall, England. The United Nations Joint Group of Experts on the Scientific Aspects of Marine Pollution (GESAMP) defined marine pollution in 1972 as "the introduction by man, directly or indirectly, of substances or energy to the marine environment (including estuaries) resulting in deleterious effects such as: harm to living resources; hazards to human health; hindrance of marine activities including fishing; impairing the quality for use of seawater and reduction of amenities." A few years later, scientists began to distinguish between marine *contamination* and *pollution,* defining contamination as "the presence of elevated concentrations of substances in the water, sediments or organisms." Contamination was seen only as a warning sign of *possible* pollution; in the rare cases where financial penalties were assessed for marine contamination, companies could consider such fines just another cost of doing business. But in most cases, waste disposal in the ocean was seen as essentially "free."

It is estimated that more than 2.5 billion metric tons of waste were generated globally in 1985. (A metric ton is about 2,204 pounds.) That figure is expected to soar to 4.5 billion metric tons by 2025, fueled by population growth, greater access to polluting technologies, and the growth of consumer societies around the world. The impact of wastes entering the ocean varies. Estuaries (the places where fresh water from rivers mixes with the salt water of the sea), harbors, and coastal waters are the most easily damaged by pollutants. The high seas may seem less susceptible to damage, but there is increasing evidence that irreversible damage already may have occurred in some areas of the open ocean and its sediments.

There are six large categories of marine pollutants: sewage, marine debris, toxic chemicals, heavy metals, oil, and radioactive materials.

Sewage

Every American flushes the toilet an average of five times each day. Multiplied by a population of about 230 million, that's more than a billion flushes every day of every month, all year long. Unless you are in a home or building that has a self-contained septic system, the waste water from flushing flows into a sewage system.

Sewage is human waste. It is "organic," meaning it comes from living organisms, and it is *biodegradable,* meaning that it decomposes with the help of bacteria. Other types of biodegradable wastes are animal excretions and plant matter, which rot where they land, ending up in the sea as runoff following rains.

Sewage can be discharged into the ocean intentionally or unintentionally. Pipes called *outfalls* often carry it to a river or directly into the sea. Sewage can also be unintentionally discharged into the sea when the capacity of a sewage system is exceeded, as often happens with older systems during heavy rains. Sewage that enters the water without any form of treatment is called raw sewage. Treated sewage has two byproducts: liquid, called effluent, and the semisolid matter that settles out of effluent during the treatment process, called sludge.

Treating Sewage

Sewage is often treated in special plants. *Primary treatment* of sewage consists of removing solids and allowing any remaining particles to settle. *Secondary* and *tertiary treatment,* also known as biological or chemical treatment, actually break down the organic matter in the sewage, which reduces the need for oxygen during the decomposition process in the sea. This makes the effluent less likely to cause eutrophication (see below). Sewage treated in this way, however, still may contain soluble chemicals and viruses. The viruses that cause polio and hepatitis A have been found to survive for months in seawater and in marine organisms; they also get into shellfish and can survive cooking.

Sludge, also known as wet cake, is produced by all sewage treatment processes. This thick, semisolid substance is dumped in landfills or coastal waters, or is mixed with effluent and discharged into the sea through pipes extending yards or miles offshore. It often contains toxic chemicals, heavy metals, harmful bacteria, and viruses, which is why it is considered unfit for use as fertilizer in the United States. Sludge disposal in the ocean was banned in the United States in 1992. There are uses for sludge; some sewage treatment plants are drying and burning sludge to produce electricity.

Much of the world's sewage is piped directly into the sea without any treatment. Although this practice is more prevalent in developing nations, it occurs in nearly every developed nation in the world—including the United States, where, for instance, raw sewage flows daily into the sea off New York City.

How Sewage Pollutes

Sewage has a relatively short lifespan—it decomposes. Microscopic organisms like bacteria and fungi satisfy their need for energy by consuming dead organic substances, a process called biodegradation. In their respiration process, however, these microorganisms consume oxygen, creating what scientists refer to as *biological oxygen demand* (BOD). If there is too little oxygen in the water to sustain these biodegraders, they die and so do all the plants and animals that depend on them. This process is called *eutrophication.* When sewage is dumped into an area without strong currents to disperse it, the oxygen is likely to be used up. Then only *anaerobic* decomposition, which needs no oxygen, can take place. Anaerobic bacteria work very slowly, and the final product smells like feces.

An extreme overgrowth of phytoplankton, called a *bloom* or *tide,* can also result when nutrient-rich sewage enters the sea. Blooms are often identified by the color of the plankton, as in a *red tide.* Blooms can clog fishes' gills. The toxins in blooms are absorbed by both fish and shellfish, and stored in their tissues.

In industrialized nations, sewage that reaches the sea generally contains industrial waste, including heavy metals and toxic chemicals. Even in less developed nations, storm water collects toxic substances from roads before entering the sea, and agricultural runoff usually contains toxic pesticides. These inorganic wastes often mix with organic matter and increase the damage done to the sea.

Monitoring Sewage

In many places around the world, sewage is not monitored on a regular basis. In the United States, the Environmental Protection Agency (EPA), the U.S. Geological Survey (USGS), the National

Oceanographic and Atmospheric Administration (NOAA), and several other federal agencies try to measure bacteria levels in fisheries, shellfish, and swimming areas. But in many cases, advocacy organizations like Baykeeper (California) and the Chesapeake Bay Foundation (Maryland) monitor coastal pollution problems and try to enforce the laws. Frequently, only when complaints are lodged is an actual bacteria count performed.

Cleaning Up Sewage Pollution

There is no way to remove or clean up sewage once it enters the ocean. The damage is done. If small amounts are dumped in an area that has strong currents, the sewage dilutes fairly quickly and is consumed by bacteria in the biodegradation process. However, the eco-balance of that area is no longer the same.

Scientists are still trying to determine the importance such changes may have over the long run. In the meantime, sludge now covers large expanses of coastal sediments, and masses of only partially biodegraded sewage are beginning to wash back toward shore, forcing the closure of many beaches, especially along the Atlantic coast of the United States.

Marine Debris

Marine debris is material, such as plastic, glass, and metal, that does not decompose or that decomposes extremely slowly. However, the term "marine debris" is now used primarily to refer to plastic litter, which is particularly hazardous because, unlike glass and metal, it floats.

Types of Marine Debris

The seabed is littered with undetonated bombs, abandoned offshore oil rigs, outdated rocket engines, drums of toxic chemicals and radioactive waste, sunken ships, automobiles, pop cans, soda bottles, pieces of wood, and other items. While this type of debris can interfere with navigation and fishing nets, plastic debris is proving to be far more hazardous, particularly to marine mammals, sea birds and fish.

Plastics have existed for only a little more than a century. Their development was sparked by the sharp rise in the price of ivory during the 1860s. Rubber shortages during World War II then stimulated the development of a wide variety of plastics and plastic products.

The plastic debris found in the ocean are of two main types: plastic resin pellets and manufactured products. Plastic resin pellets are "raw" plastic that can be melted down and molded into many types of manufactured goods. They are transported in bulk. Because they are very light, they are easily blown away by the wind. As many as 34,000 pellets per square kilometer can now be found on the surface of some areas of the Pacific Ocean.

Manufactured plastic products that become marine debris include fishing gear, shipping and packaging materials, and household plastics.

Fishing line, traps, and nets made of plastic don't rot like those made of hemp or wood, don't rust like those made of metal, and are so strong that 20-mile-wide nets are now possible. By the 1960s nearly all fishing nets were made of plastic. This has helped make fishing gear the most abundant type of manufactured plastic product in the ocean. Packaging and cargo sources of manufactured plastic products include: plastic strapping (used to bind boxes during shipment), plastic sacks, and shrink wrap.

Household products that can end up as marine debris include tampon applicators, condoms, the plastic sheeting found on disposable diapers, bags, bottles, wrappers, lids, containers, and other forms of packaging. Disposable medical equipment such as syringes is also often made of plastic and sometimes ends up as marine debris.

Many plastics take as long as 500 years to decompose. Their very strength and durability make them a persistent pollution problem.

How Marine Debris Enters the Ocean

Debris enters the ocean in many ways:

- Intentional dumping from ships
- Accidental discharge from ships or garbage barges on their way to offshore landfills
- As part of sewage
- By being swept from the land by wind or storms
- From polluted rivers and streams
- In runoff from beaches littered with trash

The Effects of Marine Debris on Living Organisms

Marine debris that sinks, such as abandoned oil rigs, old cars and the like, seems to attract fish; "artificial reefs" have even been intentionally built out of junk in some places.

Plastic marine debris, however, often floats, making it hazardous to sea birds and mammals, who either eat it or become entangled in it. Each year, more than 30,000 northern fur seals—as well as hundreds of thousands of other marine mammals, sea birds and fish—die as they become entangled in plastic bags or abandoned fishing gear.

Plastic resin pellets look like fish eggs and are eaten by sea birds and mammals; plastic grocery bags mistaken for jellyfish are eaten by sea turtles. These plastics often lodge in the intestines and stomach, blocking the digestive tract and causing malnutrition. They can also cause buoyancy that makes it impossible for marine animals to dive to escape predators.

Marine debris also interferes with navigation, and litters beaches and coastal waters. A common sight on many once-pristine beaches is a tide line of Styrofoam cups, condoms, disposable diapers, pop bottles and beer cans, shredded tires, pop tops, plastic six-pack rings, broken glass, and other by-products of civilization.

Monitoring Marine Debris

The Center for Marine Conservation is the leading U.S. organization in the field of marine debris prevention. The most extensive effort to assess the magnitude of the problem in the United States actually started as a beach cleanup effort along the Oregon coast on Saturday, October 13, 1984. Two years later, in Texas, a similar volunteer cleanup effort began to double as a pollution measurement and assessment tool. A Center for Marine Conservation worker organized a beach cleanup in which 3,000 volunteers used data cards to log the contents of the 125 tons of debris they gathered along 122 miles of beach (in just three hours!). Experts then analyzed the data, and made recommendations both on how to prevent and how to clean up marine debris. From these recommendations arose a successful program, now international in scope, called Adopt-A-Beach.

In 1988, more than 47,000 citizens from 25 coastal states and U.S. territories participated in COASTWEEKS, a program to create a baseline database on coastal debris. This data is now being used to assess the effectiveness of legislation aimed at reducing the amount of marine debris that washes ashore. Data from hundreds of similar cleanup/data collection programs that now take place each year around the world are added to the database under the auspices of the Center for Marine Conservation.

Before Annex V of the International Convention for the Prevention of Pollution from Ships, known as MARPOL, and the U.S. compliance with this international accord (the Marine Plastic Pollution Control Act, 1987), the disposal of plastics at sea within U.S. waters was legal. Although it is now possible to prosecute violators, dumping prohibitions are extremely difficult to enforce.

Cleaning Up Marine Debris

Thousands of citizen groups, school classes, clubs and organizations, and individuals participate in beach cleanup efforts along American shores each year. Divers are now beginning to lend a hand by retrieving underwater marine debris, particularly fishing gear which sometimes must be carefully extracted from coral and other living organisms. Cleanup of floating debris on the high seas is more difficult and rarely accomplished.

Solutions

Eliminating marine debris at the source, properly disposing of plastics, and developing and using photodegradable plastics (plastics that slowly degrade when exposed to sunlight), are the most effective ways to keep the ocean free of debris. In other words, prevention is a far better approach than constantly collecting thousands of tons of garbage from along our shores. Programs to actively reduce waste, including recycling and the use of degradable plastics and biodegradable nonplastic alternatives, are on the rise.

Products made of photodegradable plastic decompose as they are bombarded by ultraviolet rays in sunlight. Photodegradable six-pack rings, which have been manufactured since the 1970s, break down within about three months. In 1977, Vermont became the first state to require the use of photodegradable six-pack rings by law, and a bill mandating their use throughout the United States was introduced in Congress in 1986 and enacted in 1988. A diamond is embossed next to the finger hole on six-pack rings to indicate photodegradability. Even though these six-pack rings eventually photodegrade, it's a good idea not only to recycle them but to cut each ring as well; otherwise, it's possible they may end up in the ocean where sea birds and mammals may become entangled in them.

Laws restricting ocean dumping and tougher enforcement of such laws are needed. A ban on ocean dumping of all wastes by the year 2000 would be a great step forward. Public education is an essential element in implementing all of these measures. Additional technological solutions may also help fight the problem of marine debris.

Toxic Chemicals

We depend on chemical compounds in many ways. Thousands of chemicals have been developed to kill insects, dissolve paint, make film, and do a host of other important things. Unfortunately, many of these chemicals—toxic chemicals—are harmful to humans, plants, or animals.

Three of the most harmful chemicals are DDT, PCBs, and dioxins. DDT is a powerful insecticide, and PCBs and dioxins are primarily by-products of manufacturing processes. All three are part of a family of toxins known generically as chlorinated hydrocarbons, a large class of industrial and agricultural chemicals that do not break down easily in the environment and that tend to accumulate in the fatty tissues of animals.

DDT (dichlorodiphenyltrichloroethane) was first synthesized by a German chemist in 1874. However, the potent insecticidal properties of this chlorinated hydrocarbon were not discovered until 1939. The scientist who made this discovery earned a Nobel Prize for his work. DDT originally was used to control malaria-carrying mosquitoes; it then became widely used in general pesticides. DDT production and use began in earnest following World War II and peaked in 1963.

PCBs (polychlorinated biphenyls) are also chlorinated hydrocarbons. They are used in the manufacture of hundreds of common plastic, adhesive, and electrical products. Chemically, PCBs look so much like a form of DDT that, although they began entering the environment in 1944, they were not actually distinguished from DDT by-products until 1966.

Dioxins constitute a large family of chlorinated aromatic hydrocarbons that are all highly toxic. TCDD is considered one of the most toxic of the dioxins. Dioxins are found in wood and paper preservatives, in some herbicides, and in hexachlorophene, an antibacterial agent. They have also been found in ash and municipal incinerator smokestack effluents, and, in trace amounts, in Agent Orange, a defoliant used during the Vietnam War. Some of highest levels of dioxins in the United States are found in an estuary in South Carolina, but high levels can be found downstream of nearly every paper mill in the world.

How Toxic Chemicals Enter the Ocean

Chemicals of all kinds, including the toxic variety, enter the ocean in numerous ways.

Some DDT enters the ocean via streams and rivers as agricultural runoff. However, far more is taken up by the wind during aerial spraying of crops. This DDT in the atmosphere enters the ocean in precipitation or finds its way into the ocean via streams and rivers.

PCBs have a wide range of industrial uses because of their chemical stability. They enter the environment via paint, plastic, adhesive, and electrical equipment manufacturing processes, frequently during incineration. These compounds then find their way into the ocean in the same way as DDT. PCBs also can be dumped into rivers and then make their way to the sea. For example, two General Electric plants released 640,000 pounds of PCBs into the Hudson River in New York State during one six-year period. PCBs can also leach out of TBT antifouling paints used on pleasure boats to prevent barnacles and worms. PCBs are a hundred times more persistent than DDT residues, which means they will remain in the ecosystem for at least 500 years after entering the ocean.

Most of the dioxins in the marine environment come from untreated waste water that paper mills discharge into rivers and streams. Dioxins also enter the ocean via the atmosphere when dioxin-containing compounds such as Agent Orange are sprayed. Finally, some dioxins have entered the marine environment through intentional dumping, when toxic substances have been disposed of in the ocean.

Marine incineration is another way that all of these chlorinated hydrocarbons and other toxic chemicals enter the ocean. Huge incineration ships loaded with these wastes burn them on the high seas beyond the 200-mile limit, where few regulations and little enforcement exist. Although perhaps less environmentally damaging than simply dumping these chemicals overboard, trace metals and acidic gases are frequently released into the atmosphere during incineration. Also, the possibility always exists, whether through human error or ocean hazards, that a ship may spill its toxic cargo before incineration can take place.

Finally, dredging to deepen and widen canals and harbors disturbs sediments that may be contaminated with toxic chemicals. This can redistribute chlorinated hydrocarbons that had settled and, until they were disturbed, been kept out of the food chain.

The Effects of Toxic Chemicals on Living Organisms

Chlorinated hydrocarbons that enter the ocean are first absorbed by phytoplankton and zooplankton, microorganisms that live in the surface waters of the sea. Chlorinated hydrocarbons at concentrations lower than those often found in rainwater can prevent zooplankton eggs from developing. Entire zooplankton populations already have been decimated by these chemicals.

When the contaminated phytoplankton and zooplankton that survive are consumed by fish, the toxic chemicals lodge in the fishes' fatty tissue. Contaminated fish that are subsequently eaten by other predators pass along an even higher concentration of poisons. This process, called *bioaccumulation,* can increase the concentration of these toxic substances as much as three billion times between the one-celled plankton at the bottom of the food chain and the seals, polar bears, and humans at the top.

High concentrations of DDT cause the shells of the eggs laid by sea birds to be so thin that the chicks hatch before they can survive. Brown pelicans and double-breasted cormorants of Southern California, for instance, may be doomed to extinction, along with thousands of other types of sea birds. DDT has also been found to cause defects in the brain, spinal cord, and skeleton of animal offspring, and often causes infertility as well. Nor are sea birds the only animals so endangered—DDT residues higher than the upper limit established for bottled milk have been found in human breast milk.

Dioxins are among the most damaging of all chlorinated hydrocarbons. In humans, dioxin is thought to cause birth defects as well as rashes, nerve damage, arthritis, and liver damage. Dioxins are known to persist in our fatty tissues for decades. Even the relatively low levels of chlorinated hydrocarbons found in Arctic waters are doing a great deal of damage. Arctic mammals with thick layers of insulating, fatty tissue can accumulate dangerous concentrations of dioxin. Eskimos and other members of simple hunting societies of the Arctic who depend on these animals for food but who neither produce nor benefit from chlorinated hydrocarbons are suffering the results of very high levels of these toxins in their bodies.

Cleaning Up Toxic Chemicals

There is currently no way to clean up the persistent chemical toxins in the marine environment. They tend to accumulate both in ocean sediments and in the same 1/100th of an inch thick layer of surface water that is home to the tiny phytoplankton and zooplankton that form the base of the oceanic food chain. In some ocean areas, an Oregon researcher has found this microlayer to be contaminated with toxic chemicals in concentrations 10 to 1,000 times greater than the average for all sea water. Bioaccumulation then further increases concentrations in fish, birds, and mammals at the upper end of the food chain.

Solutions

As early as 1946 employees of the U.S. Fish and Wildlife Service warned of the extreme dangers presented by DDT, but it was not until the early 1960s that questions regarding its safety were taken seriously. Although DDT was banned in the United States in 1972, it continues to be produced and exported to other nations. Ironically, it is now known that a DDT-resistant strain of malarial mosquitoes has developed. Other pesticides and herbicides also contain chlorinated hydrocarbons. Although many are chemically similar to DDT, there are no regulations controlling their manufacture or use. Clearly, reducing the use of chlorinated hydrocarbon products is essential to reducing their presence in the ocean.

Cutting back agriculture's contribution to the chlorinated hydrocarbon load in the ocean is quite possible. For instance, organophosphates and carbamates—organic and nontoxic pesticide substitutes—already exist. Regulations requiring their use and prohibiting the use of pesticides that leave chlorinated hydrocarbon residues are needed. Mechanical pest control options, such as the disruption and removal of pests from their habitat, and greater crop diversity to establish a more complex ecosystem and encourage the existence of natural predators, are also advisable. Finally, buffer zones should be created around agricultural lands to limit runoff.

Industrial processes that generate chlorinated hydrocarbons are far more difficult to alter. However, lowering the demand for the products manufactured in this manner would certainly help, as would the installation of efficient smokestack scrubbers and the containment and treatment of all contaminated wastewater.

Heavy Metals

Heavy metals are dense elements such as lead, mercury, and cadmium that tend not to form compounds with each other, but rather with nonmetals such as oxygen and sulfur.

The most abundant metals in the earth's crust are aluminum, iron, calcium, sodium, potassium, and magnesium. Some heavy metals that have been found to be toxic are lead, mercury, cadmium, and copper. Selenium, tin, arsenic, cobalt, and manganese also occur naturally in the ocean.

Heavy metals, with the exception of mercury, are essential for life—but only extremely small quantities are needed. Anything over this small amount can be very toxic. The heavy metals that have leached into the ocean from the earth's crust over the millennia have recently been augmented by heavy metals in industrial wastes. This has led to unnaturally high levels of these toxic metals and their compounds in sea water, which has in turn resulted in banning the consumption of fish from some areas, the closure of shellfish beds, and tragedies like the one that occurred in Minamata, Japan.

How Heavy Metals Enter the Ocean

In very low concentrations, heavy metals are a natural part of sea water. But industrial processes are the main source of heavy metal pollution in the ocean. Because heavy metals are an important part of many chemical and manufacturing operations, they are mined on the ocean floor or extracted through chemical processes from seawater. Some escape during these extraction processes and add to the natural levels of metals in the ocean. On land, rain leaches cadmium from copper mine tailings; the cadmium is washed into streams and then carried to the sea. Often the tailings themselves are carried downstream, to eventually increase the level of both copper and cadmium in the ocean. Additional copper enters the ocean as it leaches out of antifouling paints used on the hulls of boats. These TBT paints have been banned in some parts of the world.

Large quantities of heavy metals enter rivers as industrial manufacturing waste and are carried downstream in the river's silt. When the silt at the mouth of the river is dredged, the resulting sludge is often dumped at sea. Heavy metals in the sludge then pollute the open ocean as they slowly drift back down to the sea floor attached to sediment particles.

Although it is difficult to accurately measure the amounts of heavy metals that enter the ocean from the atmosphere, we know that airborne particles of heavy metal compounds settle in the ocean or via rain, and that they contribute to the high levels of heavy metals found in the coastal waters of industrial areas. For example, lead, which is still used in many places as a gasoline additive, is introduced into the atmosphere in automobile exhaust, and eventually finds its way to the sea. Mercury, which enters the air during volcanic eruptions, follows a similar path. (But about one-third of the 10,000 tons of mercury used annually around the world enters rivers from chemical factories and paper pulp mills, as runoff from mercury-based pesticides, and in rain contaminated by the burning of fossil fuels.)

Finally, some metals are dumped in the ocean intentionally, as when mustard gas canisters and nerve gas rockets were dumped into the sea after World War II. In 1965, a massive amount of industrial copper sulfate was dumped off the coast of The Netherlands. Instead of dispersing, it remained in deadly concentration, killing off more than 100,000 fish, destroying many mussel beds, and decimating large areas of plankton.

The Effects of Heavy Metals on Living Organisms

Lead: Although lead may initially be dispersed when it enters the ocean, marine organisms are very good at concentrating it. This bioaccumulation can end up making fish toxic for human consumption. In waters in which the oxygen has been depleted by the biodegradation of sewage, fish build up high levels of lead even more quickly than in clean waters, adding further to the amount of lead in the food chain.

Lead poisoning in humans is now known to occur with less than half the lead exposure considered acceptable just a few years ago. Symptoms of lead poisoning include delayed development in children, nervous system disorders, learning disabilities, and mental retardation.

Mercury: Mercury is highly toxic even in very low concentrations. It disrupts central nervous system functions in animals. It is extremely easy to absorb; just breathing in organic mercury fumes can be fatal to animals and humans alike. Adding to the problem of mercury pollution in the sea is the fact that its toxicity increases when it reacts with organic pollutants like sewage and the bacteria that help sewage decompose. This reaction can produce methyl, which combines with mercury to form highly soluble methyl mercury, the poison that caused Minamata disease. High mercury levels in some large, long-lived fish like shark and tuna are making the largest specimens unfit for human consumption.

Cadmium: In humans, cadmium gravitates first to the liver and kidneys but eventually ends up replacing calcium (to which it is chemically similar) in the bones. Severe cases of cadmium poisoning, called Itai-itai, make the bones of some victims so brittle that mere coughing can cause fractures. Cadmium can also cause high blood pressure.

Copper: One of the first metals ever mined, copper is also an essential trace element required for the survival of plants and animals. Highly diluted copper salts are also powerful antibacterial agents, fungicides, and algicides. But in large doses, copper is highly toxic to animals and human beings. Chronic overexposure to copper can cause cirrhosis of the liver, stunted growth, and jaundice.

Monitoring Heavy Metal Pollution

It is now possible to measure the amounts of most of these metals in the ocean. In many parts of the world, however, no regular measurements are taken. In the United States, the EPA, USGS, NOAA, and several other federal agencies keep track of heavy metal levels in fishing areas and in waters used for recreation. But in many cases, it is left to private organizations to work to stop those who violate antipollution laws. It is often only when someone becomes sick or local citizens complain that heavy metal measurements are taken.

Cleaning Up Heavy Metals

Since background levels of heavy metals exist naturally in the ocean, cleaning up the excess contributed by human activities is virtually impossible. Until it is known what level of each is safe for marine life and humans, every effort should be made to not exceed background levels.

Solutions

Heavy metals that enter the ocean from manufacturing and mining must instead be contained and recovered. Either the value of these metals must justify such industrial recycling, or stiff fines must raise the cost of polluting the ocean, making it economically unfeasible to dispose of industrial wastes in ways that pollute.

The amount of heavy metals that enters the ocean from automobile emissions could be decreased by implementing a global ban on the use of leaded gasoline, requiring better car emissions controls, providing more mass transit, and developing and using alternative energy sources.

Oil

What we commonly refer to as "oil" is actually petroleum. Oils are organic fats, while petroleum is a thick, flammable liquid fuel considered to be of mineral origin, despite the fact that decaying plants and animals provided the initial ingredients for its formation millions of years ago.

Petroleum is composed mainly of complex combinations of hydrogen and carbon called hydrocarbons, along with organic compounds containing sulfur, nitrogen, and oxygen in varying amounts. Petroleum deposits were formed from enormous quantities of plant and animal material that built up deep layers of sediment millions of years ago. Oil drilling equipment is used to extract petroleum, or crude oil, from these ancient deposits deep below the earth's surface. The crude oil is then transported to a refinery, where it is separated, converted, or chemically treated to produce various petroleum products such as gasoline, heating oils, lubricating oils, asphalt, petrochemicals, and plastics.

Oil pollution is the most obvious kind of ocean pollution. When a tanker runs aground and millions of tons of crude oil coats the surface of the sea, the shoreline, and marine wildlife, the problem is very visible. The obviousness of oil spills made oil the main focus of anti–marine pollution efforts during the 1970s.

Although these efforts have now been broadened to deal with other types of pollutants, the damage done when petroleum products enter the ocean remains enormous.

How Oil Enters the Ocean

Nearly half of the oil in the ocean arrives there naturally through seepage from the ocean floor. The other half is contributed by oil tankers, offshore oil facilities, runoff from our roads and factories, and via the atmosphere. Some enters the sea by accident; much is deposited there intentionally.

Over its average 20-year period of production, an *offshore oil platform* can be expected to have from one to three major spills (1,000-plus barrels), at least 25 medium spills (100 to 1,000 barrels), and roughly 2,000 small spills (50 barrels). A significant amount of oil also leaks daily into the sea from these installations. *Blowouts*, which lead to very large spills, occur when extreme pressure explodes an offshore oil rig.

Ships introduce oil into the marine environment in a variety of ways, only some of which are accidental. Substandard ships and poor maintenance procedures contribute to accidents that cause oil spills, as do congested shipping lanes, reliance on faulty instrumentation, human error, and adverse weather conditions. Since supertankers can now carry millions of gallons of crude oil rather than just thousands as was the case in the past, every accident is potentially disastrous.

For many decades, oil tankers washed out their tanks and dumped oily ballast water back into the ocean while on the high seas or while at port. Although numerous conventions and laws now prohibit this practice, it is known that petroleum still enters the sea in this manner.

Rivers contribute to the ocean the oil that rains wash off our streets. Still more used oil flows into sewage systems when engine oils, for instance, are changed and dumped down sewers, or when used motor oil is used to control dust on dirt roads.

The *atmosphere* contributes the unburned portion of heating oil, gasoline, benzene and other petroleum products. These then fall with rain into the ocean.

Petroleum, however, is not the only source of hydrocarbons in the sea. Phytoplankton alone generate at least 86 billion tons of marine hydrocarbons each year, fires create ash that can fall in the sea, and plant decay also produces hydrocarbons.

The Effects of Oil Pollution on Living Organisms

The oil that pours out of a tanker that has run aground or that spews when an offshore well blows out is not spread evenly throughout the oceans as are most natural hydrocarbons in the sea. Instead, it creates highly concentrated, toxic pockets of polyaromatic hydrocarbons (PAHs).

Lighter than water, oil initially floats on the surface, soaking up sunlight and oxygen and smothering plankton. It then coats marine wildlife and sea birds. Oil destroys the natural water resistance of sea bird feathers, allowing the birds' plumage to become waterlogged and making them unable to fly. When the oil-coated birds preen themselves, they swallow oil, which causes intestinal problems and liver failure. In the northeast Atlantic alone, hundreds of thousands of oil-contaminated sea birds sink and drown every year. Those that survive oil contamination are often unable to reproduce. Also, PAHs build up in the food chain in much the same way as do chlorinated hydrocarbons and heavy metals.

It was once thought that when an oil slick began to "weather," or break up and disappear, the worst was over. But we now know this is only the beginning of a longer cycle of destruction. As oil droplets descend to the bottom of the ocean, they can clog the gills of fish and poison other sea life. Small amounts of oil pollution will "taint" shellfish (meaning they absorb enough oil to taste of it when eaten by humans, but not enough to die). Larger amounts of oil will poison or smother shellfish. As little as 2 parts per million of oil in sea water can kill lobster larvae. Once oil droplets sink into the sediment on the ocean floor, they can remain there for decades, centuries—even millennia. The natural flora and fauna of the sea die, opportunistic species that can survive in such a toxic environment move in, and the entire evolutionary process moves backwards.

Pollution of the seas by oil is also a major contributor to the deterioration of the world's coral reefs and mangrove swamps. These underwater environments have been compared to rain forests; both provide perfect habitats for a vast number of species and both are dying at unprecedented rates.

Monitoring Oil Pollution

Oil may be the most measured and monitored of any marine pollutant. While a slick remains on the surface of the sea, it usually can be monitored by aircraft and remote satellite imaging devices.

But once the slick begins to break up, it becomes far harder to track. As the oil disperses, its impact on coastlines, marine wildlife, and fisheries must be monitored. Though various U.S. agencies share the task of monitoring oil spills, many academic marine science departments participate as well. *Golob's Oil Pollution Bulletin* provides up-to-the-minute data on spills from around the globe.

Cleaning Up Oil Pollution

There are those who maintain that it is possible to clean up an oil spill. Many environmentalists, however, have come to consider the toxins too concentrated and the damage too pervasive for it to be completely mitigated. The cleanup effort following the *Exxon Valdez* oil spill in Prince William Sound, Alaska, revealed that some cleanup methods may actually do more harm than the oil itself.

Burning was used in 1967 on the slick created by the *Torrey Canyon* accident. Authorities bombarded the slick with napalm. A small fraction of the contaminants entered the atmosphere, and then fell back into the ocean when it rained. This is no longer considered an effective cleanup method.

Booms and mechanical devices, also called slick-lickers, attempt to contain a spill by surrounding it with a floating barrier, and then to remove the oil from the surface of the water by mechanical means. The effectiveness of these devices depends as much on weather conditions and the state of the seas as on the design of the equipment.

Chemical dispersants, also known as detergents, act by breaking down oil slicks into minute droplets of oil which then disperse. The chemicals used following the *Torrey Canyon* spill were found to be harmful to marine life, and the agitation needed to ensure maximum contact between the dispersant and the oil actually contaminated the water more quickly and increased the area of pollution. A family of less toxic, highly concentrated chemicals has been developed that can be sprayed directly onto the oil and that require no agitation. Improved application techniques, such as aerial spraying, are now being used.

Questions remain, however, about the impact of these chemicals on marine life. Very viscous oils, such as heavy fuel oil, cannot be treated with dispersants. Low temperatures limit the efficiency of these chemicals and weathering makes nearly any type of oil resistant to dispersion within as little as 48 hours, especially in cold climates.

Bioremediation involves using airplanes to spread nitrogen-based fertilizers on beaches affected by oil spills to speed up the consumption of oil by bacteria. This method was tried experimentally during the *Exxon Valdez* cleanup. But research crews sent to inspect the results were made ill by the fumes generated by the interaction between the nitrogen pellets and the oil.

The only completely successful method of cleaning up oil spills is to remove all the oil from the water within a very short time, and this is not yet technologically possible. Since spills occur during extraction or transportation of oil, limiting the need for oil by finding alternative, renewable sources of energy is the best long-term solution to the problem.

Solutions

Until the need for petroleum products is significantly reduced, preventing petroleum from entering the sea intentionally or by accident will continue to be necessary. A variety of approaches to this difficult task have been devised.

Intentional spillage has been reduced by changing one of the standard operating procedures on oil tankers. Storage tanks were once washed out at sea and the oily waste discharged overboard. In 1954, it became illegal to do this within 100 miles of the coast. The load-on-top system, in which the oily water resulting from tank cleaning is poured into a settling tank so that the oil can be recovered, has been used on all tankers built since 1967. This system requires that all oil terminals have facilities for accepting these waste oils. Although many laws prohibit dumping oil at sea, enforcement of these regulations is extremely difficult.

Some collisions have probably been prevented by updating and enforcing international shipping lane safety and separation regulations. When accidents have happened, often due to weather, ship malfunction, or human error, separate oil storage compartments within supertankers and double hulls have sometimes helped avoid explosions and limited oil seepage into the sea.

Reducing the demand for petroleum products could start with gasoline. Petroleum-based fuels like gasoline—known as thermochemical fuels—could be replaced in large part by biomass conversion fuels, like methane. Instead of burning fuels to create energy and producing toxins in the process, fermentation would be the primary energy conversion method. Hydrogen-based fuel cells are another potential alternative energy source that should be further developed. Solar energy, wind power, and other clean, renewable energy sources are also available.

Radioactive Materials

Radioactivity was discovered by the French physicist Antoine H. Bequerel in 1896, shortly after the discovery of X rays. Radioactive materials are the most dangerous, longest-lasting pollutants humankind is adding to our oceans. They are dangerous not only because they damage living organisms, including humans, but because some radioactive materials continue to emit dangerous levels of radiation in the environment for thousands or hundreds of thousands of years.

Radioactivity is the spontaneous disintegration of the atomic nuclei of heavy elements such as uranium, actinium, and thorium, accompanied by the emission of alpha particles, beta particles and gamma rays. These particles and rays are called radiation. Thus, radioactive materials are substances whose atoms are decaying—breaking down—into other atoms. In the process, different forms of energy—radiation—are produced. Radioactive substances can be dangerous because this radioactive energy damages living cells.

Three types of radiation concern us: *Alpha* particles are heavy particles, like those emitted by the element plutonium, that travel very quickly but not very far and can be stopped by something as flimsy as a piece of paper. *Beta* particles—electrons—emitted by radionuclides (which often form part of nuclear power plant emissions), are much smaller and light enough to penetrate human skin, but a thin sheet of lead or even aluminum can stop them. *Gamma* rays have a much shorter wavelength than sound or visible light, and are so energetic that only the thickest concrete or lead walls can stop them. Gamma rays are the most dangerous of the three types of radiation.

The rate of decay of radioactive material is referred to as its *half-life,* which is the time required for half of the material to decay. The rate of decay is constant. For example, a gram of the radioactive element plutonium, which has a half-life of 250,000 years, will have decayed to half a gram in 250,000 years, a quarter of a gram in 500,000 years, an eighth of a gram in 750,000 years, and so on. Unfortunately for living things, a sixteenth of a gram of plutonium will still be emitting dangerous radiation after one million years.

The problem, too, with some radioactive materials is that their decay products are also dangerously and persistently radioactive. A nuclear power plant, for example, "burns" one type of nuclear "fuel," but when that fuel is burned (that is, when its radiation is used to boil water, produce steam, and turn generators), radioactive by-products continue to be dangerous.

How Radioactive Materials Enter the Ocean

Radiation is everywhere; on the land, in the air, and in the ocean. Background ionizing radiation comes primarily from the continuous bombardment of the earth's atmosphere by high-energy cosmic radiation from space, and from the decay of naturally-occurring radioactive materials on the land, in the sea—even inside our bodies. In fact, every chemical element has one or more *radioisotopes,* a form of the element that has an unstable nucleus and thus emits radiation. Of the 1,000 known radioisotopes, 50 are found in nature. The rest are produced artificially as the direct products of nuclear reactions or indirectly as these products decay.

Since the 1940s, pollution of the marine environment by radioactive materials generated by humans has increased the ocean's radiation beyond its natural, "background" level—sometimes dangerously so. Several types of radioactive materials pollute the ocean.

Radioactive *fallout* was generated during explosions of nuclear weapons both during tests of such weapons and at Hiroshima and Nagasaki during World War II. Radiation from the explosions contaminates dust and other particles in the air. Eventually such particles reach the surface of the earth—they fall out of the sky. Radiation from explosions also contaminates the ground where the bomb was detonated. Contaminated soil can then wash into the sea, and contaminated dust blown into the atmosphere eventually settles over the ocean.

Land-based radioactivity in the oceans comes primarily from nuclear power plants. Nuclear power plants that use a "circulatory" cooling system sometimes discharge radioactive water into nearby waterways. Radioactive gases are discharged from nuclear power plant smokestacks and emitted from tailings left over from the milling of uranium fuel.

Nuclear power plants also produce *radioactive waste, or radwaste.* The first radioactive waste was released into the sea in 1944, when effluents were discharged from the reactors at the Hanford atomic plant into the Columbia River, which carried it to the Pacific Ocean. Radwaste is categorized as either high level or low level. A nuclear power plant generates an average of thirty tons of high-level radioactive waste each year of operation. Large quantities of both high- and low-level radwaste are also produced by nuclear weapons factories.

Nuclear-powered vehicles and nuclear-armed devices are another source of radioactive pollution. For instance, each time a nuclear submarine gets underway, the coolant system in its nuclear reactor expands and radioactive water escapes. Accidents involving nuclear-powered submarines and spacecraft, and aircraft carrying nuclear weapons, have also released dangerous amounts of plutonium into the sea.

Some scientists and nuclear critics maintain there is *no* safe level of radiation, that *all* radiation is potentially harmful—even the minute amounts once used to make clock faces glow in the dark. There is considerable controversy regarding this issue even within the scientific community.

Disposing of Radioactive Wastes

There are currently two disposal methods for *high-level radioactive waste*: it can be diluted and dispersed in a body of water (as was done at the Windscale Power Plant in Great Britain), or it can be concentrated and stored (as is done in mountain caves along the Savannah River in South Carolina). The abyssal red clay sediment that covers almost a third of the sea floor is being considered as a third alternative storage option.

Low-level radioactive waste is often released directly into rivers or into the ground. Sometimes it is mixed with concrete and solid wastes, packed into 55 gallon steel drums and dumped into the sea. Sometimes it is incinerated. Incineration, however, leaves a radioactive residue and releases radioactive particles into the air, where they can fall as rain into the sea. The U.S. Environmental Protection Agency has estimated that, should the U.S. nuclear power program continue as planned, a billion cubic feet of low-level radwaste will exist by the year 2000.

The Effects of Radiation on Living Organisms

It was more than 40 years after the discovery of radioactivity that the health hazards of radiation came to be fully recognized. It is now estimated that the human population is exposed to about twice as much radiation as it was prior to the development of artificial radioactive substances.

Radioactive substances have therapeutic uses. Extremely small doses are used to diagnose illnesses and higher doses are used to treat cancer, for instance. But any living organism can be killed by radiation if given a large enough dose. The effects of exposure to abnormally high levels of radiation can take years—even decades—to develop.

The radioactive isotopes, or radionuclides, contained in fallout that lands in the sea eventually end up in concentrated form in the bodies of marine plants and animals, and accumulate at the top of the food chain. This problem is exacerbated by the fact that humans are among the most radiosensitive of all living organisms. Many organisms near the bottom of the food chain, such as insects, can tolerate far higher levels of radiation than we can. These organisms survive and pass radionuclides in more concentrated form up the food chain.

In humans, radiation can harm both the person exposed (somatic effects) and that person's children (genetic effects). Damage to human cells is the cause of both of these types of harm. With large doses of radiation, the most immediate damage is seen on the skin (rashes, blistering, and hair loss), in the gastrointestinal tract (ulceration), and in bone marrow (infection and hemorrhaging), all places where cells divide quickly. The symptoms of *radiation sickness* usually also include nausea and vomiting.

Radionuclides are also stored in organs such as bone and bone marrow, the thyroid gland, the lungs, the liver, and the spleen, where they can continue to cause damage for years. Some types of cancer are thought to be caused by exposure to radiation.

Monitoring Radiation

The International Commission on Radiological Units and Measurements (ICRU) is the international authority on the effects of radiation. Based in Bethesda, Maryland, ICRU develops internationally acceptable recommendations regarding quantities and units of radiation and radionuclides. The International Atomic Energy Agency, located in Vienna, Austria, seeks to "accelerate and to enlarge the contribution of atomic energy to peace, health, and prosperity throughout the world," as well as establishing health and safety standards. Government agencies are usually charged with the task of monitoring levels of radiation released into the environment. Some advocacy groups do their own measurements. Greenpeace has led the movement to independently monitor, measure, and prevent marine radioactive pollution around the world.

Early measures of radioactivity that are still sometimes used include the *curie*, which expresses the number of nuclear disintegrations over a unit of time, and *rads* and *rems*, which measure the impact of radioactivity on living things. A new system of measurement uses *becquerels* (one disintegration per second) to measure the amount of disintegration activity, *grays*, (1 joule per kilogram of tissue) to measure the radiation absorption rate, and *sieverts* (dose in grays multiplied by type of particle) to measure the effective radiation dose. Radioactivity found in foods or discharged into the sea is usually measured in becquerels, while radiation limits for nuclear power plant workers are generally measured in sieverts. Radiation risk estimates are now frequently stated in terms of "ICRU Limits," rather than in terms of potential damage or in units of radioactive energy.

Scientists have developed "critical pathways" to help assess the risk of radiation to humans. The biological pathway is the food chain, in which larger animals feed on smaller organisms, passing the contamination on up the line. In the physical contact pathway, there is direct contact between a human and a radioactive object; for instance, a nuclear power plant worker picks up a tool that has been submerged in water used to dilute radioactive waste, or a fisherman handles gear that has been used in waters contaminated by radioactive waste. The atmospheric pathway starts when radiation is released into the air, for example in a discharge of radioactive gas from a nuclear power plant.

Cleaning Up Radioactive Wastes

Once radioactive particles enter the ocean, they cannot be removed or cleaned up.

Radioactive wastes that have been stored in containers and dumped in the ocean could be located and removed, but it can be argued that this in itself poses dangers. And the question would remain: what to do with it?

Solutions

Safe radioactive waste disposal methods must be developed before nuclear energy is considered a viable option. Reprocessing, a procedure that could eliminate the majority of radioactive waste, must be perfected. All nuclear testing and the use of nuclear weapons and nuclear-powered and -armed vehicles should be globally banned.

Our Changing Perceptions of Marine Pollution

Our views on marine pollution have changed as our knowledge of the ocean has changed. Historically, governments have been interested primarily in learning more about the sea to improve navigation, for commerce or for war. Accurate maps of the deep sea floor have been drawn only over the past several decades. World War II and the advent of submarines provided the impetus to study the ocean, and oceanography became a science.

Dilution, dispersion, sedimentation, and degradation are physical processes of the dynamic, self-renewing ocean that, for many years, convinced even knowledgeable scientists that this vast body of water could function as a waste receptacle. These functions were expected to adequately "digest" waste dumped in the ocean, without straining what was termed the ocean's *assimilative capacity*. The vastness of the seas helped reinforce this belief, although it was difficult to quantify what the assimilative capacity might be. But in the last several decades, scientific knowledge of the ocean's complexity has grown rapidly. During the 1950s, Soviet researchers discovered currents more than six miles below the surface of the sea. They concluded that the ocean is made up of various layers and that vertical motion exists in these layers as deep as 35,000 feet. This discovery marked the beginning of the end of the assimilative capacity theory. Something dumped in the ocean near Norway could affect people as far away as New York.

This is important because for millennia only organic matter or elements that exist naturally in the environment were introduced into the sea. Now plastics, chlorinated hydrocarbons, and refined petroleum products have been developed by humans and are not consumed by any living thing. They lie outside the biological cycle that has historically kept the environment in a state of balance. Technology has allowed humans to manipulate inorganic, often toxic, substances in large quantities, and has led to the production of unprecedentedly dangerous waste.

At the same time, scientific and technological advances over the past several decades have made it possible to monitor the impact of marine pollution on the world's ocean. This new information has led to three key concepts that shape the way we now perceive the ocean and the threats it faces: bioaccumulation, the biosphere, and environmental ethics.

Although Rachel Carson poignantly explained the process of bioaccumulation as long ago as 1962 (in her book *Silent Spring)*, the reality of the fact that marine organisms bioaccumulate and thus magnify the effects of toxic substances became abundantly clear during the 1980s. A new public understanding of this reality has made banning all ocean dumping, just a few years ago regarded as an unreasonable and extreme position, far more widely accepted. It has also led to the view now held by many environmental activists that the only effective way to stop the damage being done to the marine environment by some toxins is to stop production of these toxins altogether. The largely successful fight to stop the use of lead in gasoline is the example most often cited to support this approach to pollution prevention.

The emerging view of the earth as a single, interconnected biosphere has energized the environmental movement and promises to finally give ocean preservation high priority during the 1990s. It is now clear that the oceans mirror the health of the planet as a whole.

The most recent key concept to emerge is environmental ethics. Marine pollution, from the ethical point of view, must be judged not just in terms of costs and benefits for the current generation, but also in terms of the effects this pollution will have on the health and welfare of future generations.

Containment, rather than dispersion, is considered the best type of waste disposal methodology for most pollutants—and the ocean disperses rather than contains wastes. And the elimination of pollutants at the source, or total recuperation and recycling, is seen as the safest way to deal with the problem of marine pollution.

There is now a sense that heedless technology and careless manufacturing could make the planet uninhabitable, and that we owe it to our children and grandchildren to stop this process. A sustainable global economy, one that prevents further damage to the planet, is a logical outgrowth of this ethical position.

Conclusions

For half a century, the developed world had the technology to damage the biosphere, but not the technology to detect that damage or to undo it. For years, we were merely ignorant in our abuse of the ocean.

Now, however, it is possible to count picoplankton, the microscopic mini-phytoplankton at the base of the Earth's vast food chain. We know that as little as 25 parts per billion of DDT can limit the photosynthesis and cell division of some larger phytoplankton and that depletion of the larger phytoplankton population could damage the entire food chain. In fact, should plankton become extinct, it is quite possible that every living thing in the food chain above it would perish as well. And because phytoplankton photosynthesis plays such a vital role in using carbon dioxide in the air and producing oxygen, it is not farfetched to conclude that life on Earth could end were all phytoplankton to die.

In the scientific community and among environmentalists, the fear now exists that the next holocaust may not be nuclear, but ecological, and that it could be global in scope. Many scientists and nonscientist environmentalists alike believe that stopping this potential catastrophe must be the number one priority of the 1990s.

2

Chronology

300 The first recorded instance of marine pollution occurs in the Sea of Marmara at the mouth of the Bosphorus. It is the result of sewage from the city of Constantinople.

1680 Broad Street in New York City becomes the first public sewer in the New World.

1855 Chief Seattle, in his address to U.S. President Franklin Pierce, states, "Humankind has not woven the web of life. We are but one thread within it. Whatever we do to the web we do to ourselves. All things are bound together. All things connect. Whatever befalls the Earth befalls also the children of the Earth."

1874 A German chemist synthesizes DDT.

1886 An act of the U.S. Congress makes it unlawful to dump construction rubbish such as gravel, earth, sawdust, cinders, or mill waste into New York Harbor.

1895 The first offshore oil drilling, at what will be called the Summerland Field. Workers in southern California build wooden piers as drilling platforms.

1899 The Rivers and Harbors Act bars the "deposition of refuse" in U.S. waterways. "Refuse" is garbage and industrial waste. Sometimes called the National Refuse Act, this law contains a provision allowing raw sewage to be dumped in any body of water.

1905 The shad catch in New York Harbor, historically plentiful, plummets to less than 400,000 pounds from more than 4 million pounds in the late 1800s. This is attributed to the enormous amounts of raw sewage entering the harbor from the rapidly growing population of New York City.

1920s The engineering staff of the Los Angeles County Sanitation District decides that piping the effluent from a yet-to-be-built sewage treatment plant into the ocean will best serve the District's interests and sets to work looking for "a suitable outfall site." This decision reflects the prevailing view that the ocean can absorb almost limitless quantities of waste without suffering any damage.

1922 President Warren Harding signs a Joint Resolution of Congress requesting a conference of maritime nations to find ways to prevent the pollution of navigable waters.

1924 The U.S. Congress passes the Oil Pollution Act of 1924, making it unlawful to discharge oil into the coastal navigable waters of the United States from any vessel that uses oil as fuel or from tankers that transport oil. This is the first U.S. law to control marine oil pollution. It remains legal for ships beyond the three mile coastal limit to flush oily bilge water into the sea.

1926 After three years of study by various U.S. government departments, the International Maritime Conference convenes in Washington, D.C. to draft the first international convention on oil pollution. However, no nation agrees to ratify this document.

1929 The Swann Chemical Company invents the industrial compound PCB for use as an electrical insulator.

1930s Researchers receive the first reports of harm to sea mammals caused by marine debris—northern fur seals on the Pribilof Islands of Alaska are entangled in rubber bands cut from inner tubes, pieces of cord, string, and rawhide.

1934 Scientists discover that DDT is a potent insecticide.

The Council of the League of Nations decides to hold a conference to consider an international convention on marine oil pollution. However, Germany, Japan, and Italy refuse to attend and the conference is cancelled.

1938 Offshore oil drilling begins in the Gulf of Mexico, off the coast of Louisiana.

1939 World War II stimulates the development of plastics for use in such products as canteens and dinnerware.

1942 The first controlled, self-sustaining nuclear chain reaction takes place, marking the beginning of the atomic age.

1944 The first radionuclides are released into the marine environment when water used for cooling the reactors at the Hanford atomic plant is discharged into the Columbia River, which carries it to the Pacific Ocean.

1945 The United States detonates the first atomic bomb on July 18 at Alamogordo, New Mexico.

A U.S. bomber drops an atomic bomb on Hiroshima, Japan, on August 6. A second atomic bomb is dropped on Nagasaki, Japan on August 9.

President Harry Truman proclaims 2.4 million square miles of continental shelf along both the Atlantic and Pacific coasts to be U.S property.

1946 The Atomic Energy Commission begins dumping old oil drums packed with radioactive waste into the sea at various points along the U.S. coast.

In July, U.S military authorities conduct atomic bomb tests in the Pacific at Bikini Atoll, leaving 500,000 tons of radioactive debris in Bikini Lagoon. The inhabitants are evacuated.

1947 Britain's first atomic reactor comes into operation in August in Harwell.

Lieutenant commander George Earle IV flies secret missions to drop six metal canisters, each weighing 2 to 3 tons, into the sea 100 miles off the coast of New Jersey. The canisters contain radioactive material.

1948 The United Nations establishes the Intergovernmental Maritime Consultative Organization (IMCO) in London to focus on pollution by ships.

1951–1952 Thor Heyerdahl, the Norwegian anthropologist, explorer, and author, sails from Africa to South America on the *Kon-tiki*, a papyrus raft. He reports no sign of human pollution.

1953 The U.S. government passes the Outer Continental Shelf Lands Act, protecting the United States' jurisdiction over the seabed and subsoil of the continental shelf beyond the three-mile limit.

Inhabitants of Minamata, Japan are diagnosed with Minamata Disease, caused by eating methyl mercury-contaminated fish caught in Minamata Bay. The mercury has come from a local chemical manufacturing plant.

1954 The International Convention for the Prevention of Pollution of the Sea by Oil, known as OILPOL, is signed in Brussels. The first major convention regulating pollution on the high seas, OILPOL requires that oil and oily mixtures be emptied at port in proper facilities. The United States signs this accord.

Fish caught by Japanese fishermen are declared unfit for human consumption following a U.S. nuclear test in the Marshall Islands.

1957 The Windscale nuclear power plant in the United Kingdom malfunctions. Its number one reactor is completely destroyed, releasing more radioactivity into the atmosphere than was produced by the atomic bomb dropped on Hiroshima.

The age of plastics is heralded by two events: the opening at Disneyland of Monsanto's House of Tomorrow, in which the walls, roof, floor, rugs, and furniture are all made of plastic; and the invention of the Hula Hoop.

1958 The Geneva Conference on the Law of the Sea adopts four conventions, three of which underscore nations' responsibility to prevent marine pollution: the Convention on the High Seas, the Convention on the Continental Shelf, and the Convention on the Territorial Sea and the Contiguous Zone.

1959 In July, the First International Conference on Waste Disposal in the Marine Environment is held at the University of California at Berkeley. Scientific methods for disposing of sewage in the sea are discussed.

At the request of the Atomic Energy Commission, the National Academy of Sciences publishes *Radioactive Waste Disposal into Atlantic and Gulf Coastal Waters.* It suggests 28 carefully selected sites, each at least 75 miles apart, where low-level radioactive wastes can be safely dumped at sea. When the report is published, there is public outrage over the plan.

An International Congress of Nuclear Scientists takes place in Monaco. Oceanographers are invited to join the 300 scientists from 33 countries. Attendees are ultimately unable to decide which is safer: to bury nuclear waste in the ground or to sink it in the sea. Thus, no international control authority is established, and nuclear waste dumping in the ocean continues.

A drum containing radioactive waste washes ashore on a California beach.

1961 The U.S. Congress passes the Oil Pollution Act of 1961, in compliance with the U.S. ratification of OILPOL in 1954. It is now illegal to discharge oil into the sea in coastal zones and fisheries around the world.

1962 In January, the Ocean Transport of Radioactive Materials conference is held in New York City.

The Convention on the Liability of Operators of Nuclear Ships is signed in Brussels.

1963 In Moscow on August 5, the United States, Great Britain, and the USSR sign the Treaty Banning Nuclear Weapons Testing in the Atmosphere, in Outer Space and Under Water, known as the Test Ban Treaty. Radioactive fallout begins to decline for the first time since the end of World War II.

1964 U.S. atomic satellite SNAP-9A, containing plutonium-238, enters the Earth's atmosphere 28 miles above the Indian Ocean. By 1975, 95 percent of the plutonium will have reached the Earth as fallout, doubling the amount already present from nuclear testing.

1965 A massive amount of industrial copper sulfate is dumped just off the coast of The Netherlands. Instead of dispersing, it remains concentrated in the area where it is dumped, killing more than 100,000 fish, destroying numerous mussel beds, and wiping out plankton.

1966 The first Conference on the Status of Knowledge, Critical Research Needs and Potential Research Facilities Relating to Ecology and Pollution Problems in the Marine Environment is convened in the United States.

The U.S. Congress passes the Clean Waters Restoration Act of 1966, requiring for the first time that those who discharge oil into any navigable U.S. waters remove the pollutants or be held liable for the cost of removal.

U.S. Senator Claiborne Pell establishes the Sea Grant College program. Administered by the National Oceanographic and Atmospheric Administration (NOAA), this program provides grants to U.S. universities to study ocean engineering, aquaculture, seafood processing, coastal management, and mineral resources.

A Swedish researcher studying the effects of DDT on fish discovers the toxic nature of PCBs as well.

Researchers discover DDT compounds and PCBs in penguin eggs and in snow in Antarctica. They also devise a way to chemically distinguish PCBs from DDT.

1967 On March 18, the tanker begins to *Torrey Canyon* spills more than 100,000 tons of crude oil into waters 20 miles off the coast of Cornwall, England. This is the largest oil spill to date.

All large tankers built from this year forward must be constructed so that oily washings used for ballast can be stored in a slop tank and not washed into the sea.

1968 A U.S. bomber loaded with atomic weapons crashes into the ocean near Thule, Greenland. High concentrations of plutonium are found as far as 9 miles away.

The National Multi-Agency Oil and Hazardous Materials Contingency Plan is released in the United States, one of many responses to the 1967 *Torrey Canyon* spill.

1969 On February 18, the President's Panel on Oil Spills meets for the first time.

On February 24, Union Oil's number A-41 well blows out of control in the Santa Barbara channel, just off the coast of

California. As many as 21,000 gallons of oil escape every day for months, creating a 1,200 sq km oil slick in the Pacific before the well is capped.

The U.S. Congress passes the National Environmental Policy Act (NEPA) of 1969, the most comprehensive and far-reaching environmental law in the United States NEPA requires that environmental impact statements be prepared for all significant federal projects.

The International Convention on Civil Liability for Oil Pollution Damage, known as the CLC, is drafted by the Inter-Governmental Maritime Consultative Organization (IMCO) in response to the *Torrey Canyon* disaster. It provides the legal framework that enables governments and other parties damaged by marine oil pollution to recover oil spill cleanup costs and damages. At the same time, the Convention Relating to Intervention on the High Seas in Cases of Oil Pollution Casualties is drafted.

TOVALOP (Tanker Owners Voluntary Agreement concerning Liability for Oil Pollution) is enacted on September 18 by the International Tanker Owners Pollution Federation (ITOPF). TOVALOP, a worldwide agreement to meet damage and cleanup costs resulting from oil tanker accidents, provides compensation to governments and tanker owners in the event of an oil spill.

Fifty nations sign the International Convention Relating to Intervention on the High Seas in Cases of Oil Pollution Casualties.

During the annual commercial seal hunt, U.S. fur seal managers begin to monitor the incidence of seals becoming entangled in marine debris.

The annual world fish catch declines for the first time in 25 years.

1970 Seven Chevron wells off the coast of Louisiana catch fire in February. Approximately 1,000 barrels of oil leak into in the Gulf of Mexico each day from February 10 until March 31, causing a large slick.

The U.S. Congress passes the Water Quality Improvement Act, which makes it possible for the first time to fine or

1970 (*cont.*) imprison someone for failing to report an oil spill or for knowingly discharging oil into certain types of waters. It also authorizes the President to take any action needed to clean up a spill and to charge the responsible party, and establishes a federal revolving fund for cleanups.

The U.S. government convenes the Study of Critical Environmental Problems (SCEP) to generate information for the United Nations Conference on the Law of the Sea.

The United Nations Food and Agriculture Organization (FAO) sponsors the Technical Conference on Marine Pollution and Its Effects on Living Resources and Fishing in Rome.

Excessive amounts of methyl mercury are found in five cans of tuna fish destined for sale in the United States, leading to the recall of one million cans of tuna.

In August, the U.S. Army launches Operation CHASE, sinking a ship loaded with nerve gas canisters in the sea between Florida and the Bahamas. Protests by U.S. citizens and the Caribbean nations go unheeded.

1971 An international scandal occurs as a Dutch ship, the *Stella Maris,* attempts to dump canisters containing 650 tons of chlorinated hydrocarbons from The Netherlands off the coast of Norway in the North Sea. The ship then heads toward an area 1,000 miles off the coasts of Ireland and Iceland, where the United States frequently dumps wastes, but angry fishermen refuse to refuel the ship. It is forced to return to Rotterdam, where its cargo is unloaded.

Thirty-four nations sign the International Convention on the Establishment of an International Fund for Oil Pollution Damage.

The United Nations Food and Agriculture Organization (FAO) convenes its Sixteenth Governing Conference in November. The controversy regarding the use of DDT peaks at this conference.

Bikini Atoll is declared safe, but after those who move back experience a 75 percent increase in contamination by caesium-137, a radioisotope, the area is deemed uninhabitable for another 50 years.

1972 The U.S. Congress passes four major environmental protection laws: the Ports and Waterways Safety Act, the Federal Water Pollution Control Act Amendments, the Marine Protection, Research, and Sanctuaries Act (also known as the Ocean Dumping Act) and the Coastal Zone Management Act (CZMA).

Sixty-one nations sign the International Convention for the Prevention of Marine Pollution by Dumping of Wastes and Other Matter, also known as the Ocean Dumping Convention. This convention is considered the first international treaty devoted to global environmental protection.

The first United Nations Conference on the Law of the Sea (LOSC) is held in New York. Ten annual sessions of this historic conference and arduous negotiations will be needed to complete the accord.

Ninety-five nations agree to abide by the International Collision Regulations. These rules promise to help prevent collisions at sea.

The United States bans the dumping of radioactive waste into the ocean.

Thor Heyerdahl, the Norwegian anthropologist, makes his second journey from Africa to South America. This time he observes drifting clots of oil floating on the surface on 40 of the 57 days he is at sea.

1973 The satellite Cosmos 954 fails to achieve a higher orbit and enters the earth's atmosphere on January 24. Radioactivity from the satellite poisons the atmosphere and radioactive debris contaminates an area in northern Canada. President Jimmy Carter pledges to pursue a ban on nuclear power in space but later withdraws the proposal.

The International Convention for the Prevention of Pollution from Ships, known as MARPOL, is ratified. This convention supersedes the 1954 International Convention for the Prevention of Pollution of the Sea by Oil. This landmark accord has broader coverage and tougher enforcement provisions than any previous international agreement.

International Atomic Energy Agency Regulations for the Safe Transport of Radioactive Materials are issued.

1973 (*cont.*) The Marine Protection, Research, and Sanctuaries Act of 1972 takes effect on April 23, marking the end of unregulated disposal of waste material in U.S. coastal waters.

Twenty nations sign the protocol Relating to Intervention on the High Seas in Cases of Marine Pollution by Substances Other than Oil. This protocol reveals the growing awareness that more than just oil threatens the health of the oceans.

1974 The Paris Convention, the first international agreement to deal with land-based marine pollution, goes into effect, regulating marine pollution from land-based sources throughout the European Community. "Blacklisted" substances include: chlorinated hydrocarbons, other halogenated organic compounds, mercury and cadmium compounds, plastics, oil, and petroleum-derived hydrocarbons.

1975 The Finnish tanker *Enskeri* is loaded with 690 barrels (8 tons) of highly toxic arsenic wastes and heads for a dumping site in the south Atlantic. Protests from the governments of the United States, Brazil, and South Africa, and other nations lead to United Nations and Organization of American States meetings but fail to prevent the dumping.

Recognizing the harm caused by land-based marine pollution, the United Nations Economic Commission for Europe sponsors a conference on the Protection of Coastal Waters Against Pollution from Land-Based Sources.

The International Atomic Energy Commission defines high-level radioactive waste. The IAEC sharply decreases the allowable radioactivity in radwaste.

1976 Sewage fouls Long Island beaches and causes massive fish kills along the coastline of the mid-Atlantic states.

On December 15, the *Argo Merchant* runs aground on the Nantucket Shoals, spilling its entire load of 28,000 tons of dense fuel oil before salvage attempts end. This incident prompts President Jimmy Carter to call for international action and spurs amendments to existing legislation.

The Toxic Substances Control Act bans PCB production in the United States.

1977 In April, a well in the Ekofisk field in the Norwegian sector of the North Sea blows out. Ironically, this occurs as blowout preventers are being installed. About 30,000 tons of oil escape during the eight days required to cap the well.

The Nature Conservancy Council sponsors an International Meeting on Wildlife and Oil Pollution in the North Sea.

International Regulations for Preventing Collisions at Sea, which separate ship traffic into one-way lanes to reduce the risk of collision (and consequent pollution) in congested waters, are ratified.

At an approved site in the northeastern Atlantic 400 miles off the European coast, the United Kingdom dumps 2,250 tons of radioactive waste packed in tarred, concrete-encased steel drums.

The Netherlands and Switzerland dump 4,180 containers of radioactive waste in the ocean.

1978 The U.S. Environmental Protection Agency sues the city of Philadelphia for ocean dumping violations. Philadelphia switches to land disposal.

A hydraulic pipe in the steering engine of the *Amoco Cadiz,* a 233,000-ton Liberian tanker, breaks on March 16 and the ship runs aground just 1 1/2 nautical miles off the coast of Brittany, France. More than 90 percent of its full cargo of oil spills into the sea over a period of eleven days. The tanker is bombed at the beginning of April to release the remaining oil. More than 80,000 tons of weathered oil (called "chocolate mousse") washes ashore, causing dizziness, headaches, and nausea among coastal residents.

In May, the Greek tanker *Eleni V* breaks in half after colliding with a French ore carrier off the coast of Norfolk, England. The stern is towed to shore, but the bow continues to leak large quantities of oil for almost a month before the Royal Navy bombs it, creating a 300-mile oil slick and additional tarred beaches.

In October, the Greek tanker *Christos Bitas* goes aground off the coast of northern Europe. About 3,000 tons of oil are transferred to other ships, but the remaining 1,000 goes down with the vessel when it is intentionally sunk off the southern coast of Ireland.

1978 (*cont.*) The 190,000-ton *Esso Bernicia* collides with a jetty at an oil terminal in the Shetland Islands off the coast of Scotland in December. More than 1,100 tons of heavy oil despoils 30 miles of coast and kills many marine birds.

1979 In January, the tanker *Betelgeuse* explodes at a Gulf Oil Company offshore terminal along the southwest coast of Ireland. Fifty-one people are killed and more than 60 miles of coastline is polluted. Gulf is criticized for poor safety standards and for attempting to cover up its involvement in the incident.

CRISTAL (Contract Regarding an Interim Supplement to Tanker Liability for Oil Pollution) takes force in April. This voluntary agreement among oil companies provides supplemental compensation for all oil pollution damage beyond that which is legally obtainable elsewhere.

In June, an offshore oil well in the Ixtoc field off the Mexican coast blows out, spewing at least 500,000 tons of crude oil into the Gulf of Mexico in the largest blowout to date. It takes nine months to bring the ensuing fire under control.

Two tankers, the *Atlantic Empress* and the *Aegean Captain,* collide near Tobago in July. Twenty-six seamen die and 90,000 tons of oil spill. More than 210,000 additional tons of oil sit in a corroding, sunken tank at the bottom of the Atlantic.

In December, a ship loses 51 cylinders of chlorine gas off the Dutch coast; 15 cylinders are never recovered.

The plastic grocery bag is introduced.

1980 Oil drilled offshore accounts for more than 20 percent of the total world production, or almost 14 million barrels per day.

The Funiwa-5 well off the shore of Nigeria blows on January 17, spilling about 27,000 tons of oil before it caps itself after two weeks. The drinking water and food supply for 250,000 people are contaminated and nearby mangrove swamps are severely damaged.

On October 2, the Hasbah-6 well off the Saudi Arabian coast blows out, killing 19 workers. By the time it is capped a week later, 5,000 to 13,000 tons of heavy crude oil create an enormous slick that damages the coasts of Bahrain and Qatar.

1981 Judge Abraham Soafer of the U.S. District Court rules in favor of New York City, which had sued the EPA to avoid having to comply with the 1981 deadline to stop all ocean waste dumping. Both parties ignore the London Convention; ocean dumping not only continues but increases. Hundreds of U.S. municipalities apply for ocean dumping permits.

1982 The United Nations holds its final session on the Law of the Sea Convention (LOSC). The Convention contains 320 articles and nine annexes, representing nearly a decade of painstaking effort to devise an international legal foundation for maritime law, marine pollution, marine research, and control of the deep seabed. Although the United States takes a leading role in the negotiations, President Ronald Reagan refuses to sign the treaty.

Representatives from 14 North European nations sign the Port State Control Memorandum of Understanding in Paris in July, establishing shipping safety reporting mechanisms and an international data bank.

France becomes the first nation in the world to ban the use of tributyltin (TBT) antifouling paints on the hulls of boats.

To help combat marine pollution, Sweden limits the use of the heavy metal cadmium.

1983 The Environmental Protection Agency and the governments of Virginia, Maryland, Pennsylvania, and the District of Columbia approve the Cooperative Chesapeake Bay CleanUp Agreement.

James Thornton, a Natural Resources Defense Council attorney, asks the Bethlehem Steel Corporation to stop dumping toxic chemicals into the Chesapeake Bay and threatens to sue if the company refuses. This marks the beginning of the "citizen's enforcement" movement, in which citizens try to force companies to comply with pollution laws already on the books.

Iraq bombs two Iranian offshore oil wells. More than 2,000 barrels of oil a day begin spilling into the Persian Gulf. Nine more wells are bombed by the end of the year.

Greenpeace divers discover high levels of radioactivity in the Irish Sea off the coast of Sellafield, Cumbria, near a pipeline

1983 (*cont.*) used by British Nuclear Fuels (BNFL). BNFL prevents Greenpeace divers from plugging the pipe and obtains a legal injunction against the organization. A judge fines the group £50,000. Contaminated mussels are found and the Department of Energy closes miles of beach.

1984 In January, in the seas between Denmark and Britain (and dangerously close to a field of offshore oil rigs), the Danish cargo ship *Dana Optima* runs into heavy weather. Eighty drums, containing 16 tons of the highly toxic weedkiller Dinoseb (dinitrophenol), are washed overboard. Searchers fail to locate the drums.

The first Workshop on the Fate and Impact of Marine Debris is held in Honolulu to identify the scientific and technical aspects of the marine debris problem and its impact on marine species. This is the first time concerned individuals meet to share and discuss observations of the impacts of marine debris.

In August, the French cargo ship *Mont Louis,* carrying 200 tons of uranium hexafluoride in 30 heavy steel containers, collides with a cross-channel ferry and sinks in 46 feet of water 11 miles off the Belgian coast. All the containers are eventually recovered intact, but grave risks are run during the recovery operation and there is a great deal of misinformation and secrecy regarding the nature of the shipment.

The first Beach Clean-Up takes place along the Oregon coast on Saturday, October 13. The Center for Marine Conservation soon turns this event into an international effort, logging the types and amounts of debris picked up by tens of thousands of volunteers along coastlines around the world.

A ship loses 2,500 drums of the volatile chemical methyl ethyl ketone (MEK) in the Bay of Biscay off the coast of England in October. In November, a 45-gallon drum of the MEK explodes on a Dorset, England beach. The next day, two more drums of the substance explode on a Portsmouth beach. A total of 22 drums wash ashore.

The Sierra Club Legal Defense Fund files a lawsuit against Union Oil (Unocal), accusing the refinery of illegally dumping millions of gallons of waste water contaminated with toxic chemicals and oil into San Francisco Bay. Tom Billecci, a former Unocal employee, plays a key role in this landmark Sierra Club Legal Defense Fund (CLDF) law suit (see chapter 3).

The First North Sea Conference is held in Bremen, West Germany. The severe pollution of the North Sea is recognized and possible solutions to the problem are discussed.

Dr. John V. Byrne, head of the National Oceanographic and Atmospheric Administration (NOAA), declares 1984 "The Year of the Ocean" on the first anniversary of the U.S. "acquisition" of the continental shelf. The Year of the Ocean is a forum for development, rather than an environmental opportunity, despite protests by the Oceanic Society, among others.

1985 A young sperm whale is found dying along the New Jersey shore. A mylar balloon is found lodged in its stomach and three feet of purple ribbon wind through its intestines.

The Greek ship *Ariadne* goes aground in the harbor of Mogadishu, Somalia, in August. Highly toxic solvents, pesticides, and tetra-ethyl lead contaminate the harbor, forcing the evacuation of 300,000 people and shutting down all waterfront businesses. A salvage barge does not arrive until two weeks after the accident.

1986 In October, a Soviet submarine catches fire off the coast of Bermuda. The sub, reportedly powered by two nuclear reactors and capable of carrying 16 nuclear missiles, is intentionally sunk about 700 miles from Bermuda.

1987 The European Year of the Environment.

The U.S. Congress passes the Marine Plastic Pollution Control Act of 1987, outlawing the disposal of plastics at sea within U.S. waters.

The Clean Water Act is enacted over President Ronald Reagan's veto. It provides for sewer construction and the cleanup of estuaries, toxic waters, and polluted runoff.

The Eros 2000 Program begins in July, providing a forum through which the European Community can begin to understand the ocean and prevent coastal pollution.

The Second North Sea Conference is held in London. Signatory nations agree to reduce by half the amount of phosphates and nitrogen dumped in the North Sea; permit the dumping of wastes at sea only if it can be demonstrated that there are no

1987 (*cont.*) practical alternatives on land and that the substances dumped will not damage the marine environment; and end the incineration of chemical wastes at sea.

Both houses of the U.S. Congress pass legislation to restrict the use of TBT-based antifouling paints commonly used on pleasure boats to prevent barnacles. TBT is the first chemical to be regulated on the basis of its environmental impact rather than on the basis of its potential effect on human health.

Tons of medical and other wastes wash up on beaches along the U.S. Atlantic coast, focusing media attention on marine pollution.

Fishermen from the North Pacific Rim sponsor an international conference in Kona, Hawaii on marine debris. Fishing vessel activity guidelines are adopted.

1988 The Environmental Defense Fund releases a report in April stating that acid rain accounts for 25 percent of the nitrogen in Chesapeake Bay, as much as is contributed by raw sewage and industrial plants combined. The report is timed to influence a meeting between U.S. President Reagan and Canadian Prime Minister Brian Mulroney.

Amoco Oil Company is ordered to pay French victims of the marine pollution resulting from the *Amoco Cadiz* oil spill of 1977 a record $85.2 million in damages as partial compensation for the $2 billion cleanup costs.

More than 47,000 citizens from 25 coastal states and U.S. territories participate in COASTWEEKS; they compile a baseline database on coastal debris under the auspices of the Center for Marine Conservation in Washington, D.C.

In Sylt, West Germany, 30,000 people form a 25-mile human chain to protest the pollution of the North Sea.

One hundred thousand Estonians, Lithuanians, and Latvians link hands along the Baltic Sea to call for a cleaner environment.

On December 31, the U.S. Senate ratifies Annex V of the International Convention for the Prevention of Pollution from Ships. Military vessels are exempted until 1992.

1989 On March 24, the *Exxon Valdez* runs aground, spilling 11 million gallons of crude oil in the pristine waters of Prince William Sound off the coast of Alaska. The impact of the largest oil spill to date in the United States is similar to the international impact of the *Torrey Canyon* disaster.

On April 28, the International Maritime Organization agrees on new operating procedures to streamline the management of environmental disasters at sea. These procedures involve bonuses for salvage operators able to prevent unnecessary environmental damage and authorize the master of a stricken vessel to resolve the salvage price without referring the matter back to the ship's owner.

The Iranian oil tanker *Kharg-5* is ripped by an explosion on December 19 while moving north of Las Palmas in the Canary Islands. The ship is abandoned after fire breaks out; the 35 crew members are picked up by a Soviet freighter. Attempts to tow the stricken vessel fail. The 230-square-mile slick created by the tanker's lost cargo of 37 million gallons of crude oil affects the coastal waters off Morocco.

1990 On January 2, a major oil spill from an underwater pipeline carrying heating oil from Exxon Corporation's Bayway Refinery to its plant in Bayonne, New Jersey is discovered. As much as 500,000 gallons of oil creates a 20-square-mile slick (contained within booms). Dead birds are found in nearby marshes. An additional 24,000 gallons of heating oil leaks from a hole in a barge on February 28, and 3,500 gallons leak from a barge at Exxon's Bayway plant on March 1. Exxon comes under heavy criticism and investigations are planned.

The U.S. Congress passes the Oil Pollution Act by unanimous vote in both houses on August 4, following fifteen years of discussion. The Act covers liability, cleanup, and prevention of oil pollution, and phases out use of single-hull tankers by 2010.

The U.S. Congress approves the first significant rewrite of the Coastal Zone Management Act (CZMA) since its original passage in 1972.

On June 7, the British tanker *BT Nautilus* runs aground in New York Harbor, spilling 260,000 gallons of heavy crude oil while attempting to dock at the Coastal Oil terminal in

1990 (*cont.*) Bayonne, New Jersey. The spill damages New York and New Jersey coastlines; thick tar balls wash up on New York beaches.

On July 28, two barges carrying partly refined oil collide with the Greek tanker *Shinoussa* in Galveston Bay, Texas, spilling 500,000 gallons of heavy crude oil into the bay. Galveston Bay is declared a disaster area and the Texas Department of Health halts all fishing, shrimping, and oystering in the vicinity of the spill. The owner of the barges, Apex Towing, announces on August 2 that it is no longer paying for the cleanup; the U.S. Coast Guard threatens to pursue the matter in court.

The world's largest industrial nations agree November 1 to phase out the dumping of industrial waste at sea by 1995. A survey is planned to identify the areas most heavily polluted by dumping. The agreement calls for greater recycling efforts, cleaner industrial processes, improved treatment of wastes, and research and development of alternative and environmentally sound methods of waste disposal.

After five years in court, the Sierra Club Legal Defense Fund wins an out-of-court settlement in its landmark lawsuit against Union Oil (Unocal) for illegally dumping millions of gallons of waste water contaminated with toxic chemicals and oil into San Francisco Bay. Unocal is ordered to pay a precedent-setting $4.22 million penalty.

1991 Iraq destroys several hundred Kuwaiti oil wells during the Persian Gulf War. Millions of gallons of oil spill into the Gulf, making this the largest marine oil disaster in history.

Alaska implements the toughest oil spill regulations in the nation. Based on the Oil Pollution Act of 1990, Alaska requires oil pipeline and shipping companies to have enough equipment on hand to contain or clean up 60 percent of a tanker's cargo within three days of a spill. Incentives are simultaneously offered to voluntarily impose stringent spill prevention measures, like double-hulled tankers and escort ships. The regulations go into effect amidst complaints from experts that 100 percent cleanup should be required.

The Exxon Corporation, owner of the *Exxon Valdez,* agrees in March to pay $1 billion in damages and fines over a seven-year period.

In June, eight "circumpolar nations"—Canada, the United States, Iceland, Denmark, Norway, Sweden, Finland and the Soviet Union—sign the first of a series of ground-breaking multilateral protocols on Arctic pollution.

On June 24, the U.S. Environmental Protection Agency announces that 17 acres of industrially contaminated marine sediments in Commencement Bay in Tacoma, Washington have been cleaned up. This project constitutes the first completed shoreline pollution cleanup of the Superfund program.

A settlement is reached on July 11 on a Justice Department lawsuit against the state of Florida over the pollution of parts of the Florida Everglades by large sugar and vegetable farms to the north. The suit costs the state at least $6 million in legal fees. The farmers are required to cut discharges by 25 percent by the year 2002, and the state agrees to create a large artificial marsh to act as water filter between the farms and the Everglades.

On September 30, 2,000 gallons of radioactive coolant water escape through valves into the water system used to clean equipment at the Seabrook, New Hampshire, nuclear power plant. Four plant workers are contaminated and at least 45,000 gallons of radioactive water is discharged a little over a mile offshore through the plant's normal cooling water discharge.

1992 The U.S. Navy must comply with Annex V of the International Convention for the Prevention of Pollution from Ships. Military vessels were exempted when the convention was ratified by the United States in 1988.

The United Nations Earth Summit (UNSET) takes place in Rio de Janeiro, Brazil, on July 3 through 14. President George Bush is a lone holdout on many of the environmental agreements proposed during this important conference.

On July 15, Alaska Governor Walter J. Hickel vetoes a $50 million legislative package for cleanup projects related to the 1989 *Exxon Valdez* oil spill. The measure, which would have been funded by Exxon settlement money, was passed by the state legislature and supported by environmentalists.

The harbor at New Bedford, Massachusetts is reported to have a higher underwater concentration of PCBs than any other

1992 (*cont.*) site in the nation. Two companies accused of contaminating the harbor agree to pay $21 million in fines and restitution. This adds to the $90 million collected through settlements with other harbor polluters.

Democratic presidential candidate William J. Clinton and running mate Al Gore win the 1992 U.S. presidential elections. Al Gore is considered an environmentalist and is expected to push for the reversal of many of the environmental policies of the preceding Republican administrations.

In November, a Japanese freighter picks up 1.7 tons of plutonium from a French nuclear fuel reprocessing plant. The plutonium is to be transported to Japan for storage and ultimately for use in nuclear power production. Greenpeace and other environmental groups charge the fuel could be hijacked or that the ship could catch fire and contaminate the ocean. Attempts to stop the shipment fail.

In November, the Union of Concerned Scientists issues the "World Scientists' Warning to Humanity," which declares that the "destructive pressure on the oceans is severe." It calls for fundamental changes in the levels of consumption and the way of life in industrialized nations, and for an end to environmentally damaging activities.

A Greek tanker, the *Aegean Sea,* runs aground in shallow waters off the coast of Spain on December 3, during heavy weather. The ship cracks open in choppy seas, spilling 23 million gallons of oil. Rich fishing grounds are threatened and thousands of coastal residents are evacuated.

1993 On January 5, the U.S.-owned oil tanker *Braer* is driven aground in a heavy storm off the southern tip of the Shetland Islands, 100 miles northeast of Scotland in the North Sea. Within a week, hurricane force winds shatter the ship's hull, spilling its cargo of 26 million gallons of crude oil. The Scottish government bans all fishing within 50 miles of the wreck, more than 6,000 sea birds die, and inland sheep pastures and gardens are contaminated by oily residues.

On January 21, the *Maersk Navigator,* a fully-laden Danish supertanker, collides with the *Sanko Honour,* an empty Japanese tanker. More than 25,000 tons of light crude oil spill into the Andaman Sea, off the coast of the Indian island of Sumatra.

The *Maersk Navigator* catches fire and burns for five days. Booms and detergents fail to contain the spill, which threatens the rich marine life of nearby archipelagoes.

Scientists from the National Oceanographic and Atmospheric Administration (NOAA) denounce the findings of an Exxon Corp. study of the long-term environmental damage caused by the *Exxon Valdez* oil spill. The NOAA scientists dispute Exxon's claims that up to 90 percent of the affected shores had recuperated, accusing the corporation of "censoring" information and misinterpreting data. The government scientists state that the spill has done lasting damage to the ecosystem.

In July, the National Resources Defense Council (NRDC) releases its third annual report on beach water quality monitoring and beach pollution: *Testing the Waters III: Closings, Costs and Cleanup at U.S. Beaches.* On more than 2,600 occasions in 1992, beaches were closed or advisories were issued against swimming in the 22 states which monitor beach water quality. This represents an increase over 1991. High levels of bacteria account for the overwhelming majority of beach closings. The NRDC calls for a strengthened Clean Water Act.

Leading Mexican environmentalists accuse the U.S. government of covering up a 4,000-ton spill of highly corrosive sulfuric acid into the ocean of Mexico's Pacific coast. The charges are made as environmental concerns delay negotiations of the North American Free Trade Agreement (NAFTA).

3

Biographical Sketches

MARINE POLLUTION IS A MODERN PHENOMENON and preservation of the oceans is a recent concept. Because of this, a number of the people who have most profoundly influenced our current views on the sea have been environmentalists with no special training in the marine sciences. Barry Commoner and Helen Caldicott fall into this category.

Others are ocean scientists who have made major contributions to our understanding of the marine environment but did not necessarily oppose including the sea in waste management programs. Bostwick Ketchum and Edward Goldberg are two such scientists.

Still another category includes activists, some of whom are scientists and some not. Although not all of them are as well known as Jacques-Yves Cousteau, each one mentioned in this chapter has played a key role in fighting a particular type of marine pollution or in protecting a particular area. These activists include Terry Backer and Tom Billecci (illegal dumping), Kathryn O'Hara (marine debris), Marjory Stoneman Douglas (Florida Everglades), and Michael Herz (San Francisco Bay).

Each of these individuals has left a unique set of footprints in the sand.

Terry Backer

Connecticut oysterman Terry Backer founded the Connecticut Fishermen's Association to help fight violations of state and federal antipollution laws that were threatening the state's oyster beds. Shellfish are an $8 million-a-year business in Connecticut, but the even larger real estate boom has overloaded sewage systems, causing bacterial contamination that periodically shuts down the state's oyster beds. The Fishermen's Association won an out-of-court settlement with the City of Norwalk over waste treatment plant violations and fought to save a tidal mud flat in Norwalk Harbor. Backer then became the head of Long Island Keeper, an organization that tried to prevent Long Island Sound from becoming even more polluted. "Don't call me an environmentalist—I hate that," he is quoted as saying in 1987. "I'm a businessman—and my resource is being threatened." The recognition that economic survival depends, not on less stringent pollution regulations, but on the preservation of natural resources, distinguishes Backer as a pioneer in the forefront of a new era for environmentalism.

Tom Billecci

While working for Union Oil of California (Unocal), Tom Billecci discovered that his employer was polluting San Pablo and San Francisco Bays. He tried to alert the EPA, the California Regional Water Quality Control Board, and several other agencies to the violations, but didn't succeed. Finally, with the help of the Sierra Club Legal Defense Fund, Billecci brought suit against his former employer in 1984. Billecci became a key Sierra Club Legal Defense Fund witness, and the citizen's suit provisions of the Clean Water Act as amended in 1972 allowed Billecci to face off in court against Unocal. The case went all the way to the Supreme Court. Unocal finally agreed in 1990 to pay a $5.5 million out-of-court settlement, half of which was given to the Trust for Public Land for the acquisition and restoration of wetlands in the San Francisco Bay area. Unocal has now installed a $64 million wastewater treatment facility at the refinery, and other firms in the area have upgraded their facilities as well. Billecci proved that one person can stand up to a multinational corporation and win. By daring to "blow the whistle" on an employer who violated marine pollution regulations, he serves as a role model for others.

Gro Harlem Bruntland

The prime minister of Norway, Gro Harlem Bruntland is also one of the most outspoken world leaders on ecological matters. She wrote an influential report for the United Nation's General

Assembly, *Our Common Future* (1987, World Commission on Environment and Development). The report explained and popularized the concept of "sustainable development," or development that meets the needs of the present without compromising the ability of future generations to meet their own needs. Many governments and political leaders have found this concept to be a powerful rallying point, and a more palatable ecological stance than the no-growth, eliminate-all-toxins approach advocated by some environmentalists. Environmental gatherings saw a surge in attendance following Ms. Bruntland's history-making report, and the boost provided by this report to the environmental movement cannot be exaggerated. Although sustainable development is not a concept unanimously accepted by environmentalists, it has provided a middle ground where discussion can finally take place. Ms. Bruntland was also instrumental in founding The Centre for Our Common Future, a Swiss-based organization whose aim is to sustain the momentum generated by *Our Common Future* and to create a broad-based international environmental movement. This organization has been active in laying the foundation for several international conferences on the environment.

Helen Caldicott

Born in 1938, Helen Caldicott, a native of Australia, is a pediatrician and one of the world's foremost antinuclear activists. Recognizing the severe health dangers of radioactive contamination, she led successful campaigns during the early 1970s to ban French atmospheric nuclear testing in the South Pacific and the export of uranium by Australia. She published *Nuclear Madness: What You Can Do* in 1978 and single-handedly led a successful protest in Australia that forced France to finally halt nuclear testing in the Pacific Basin. Inspired by Bertrand Russell's effort to rid the earth of all nuclear weapons, Dr. Caldicott has dedicated her life to preventing the contamination of the planet by radioactivity.

Dr. Caldicott is also known for her work with children born with cystic fibrosis, and she has taught pediatrics at Harvard Medical School. Although the Physicians for Social Responsibility was founded in 1962, she revitalized it in 1978 and made it a driving force in the antinuclear movement in 1979 following the Three Mile Island nuclear power reactor crisis. Dr. Caldicott also founded, in 1980, the Women's Action for Nuclear Disarmament, a Washington lobbying group. She is a giant in the field and may have done more to protect the marine environment from radioactive contamination than any other single individual.

Rachel Carson

Rachel Carson (1907–1964), considered by many to be the mother of the environmental movement in the United States, was drawn into zoology by her love of the outdoors. She earned bachelors and masters degree in zoology and did postgraduate studies in genetics and marine biology, the latter at Woods Hole Oceanographic Institution. She then began working for the U.S. Fish and Wildlife Service. Her book *The Sea Around Us* (1951) thrust her into national prominence. An eloquent treatise on the sea, the book made little mention of the threat of pollution. Ms. Carson, however, became increasingly aware of the threat of chlorinated hydrocarbons in pesticides and other toxins during the late 1950s. In *Silent Spring* (1962), she described the bioaccumulation of DDT and other toxins in the food chain, from plankton to fish to birds. In this masterpiece that shocked the nation she wrote: "For all at last returns to the sea—the beginning and the end."

Sarah Chasis

A senior attorney with the Natural Resources Defense Council (NRDC), Sarah Chasis was named Coastal Steward of the Year by the National Oceanographic and Atmospheric Administration (NOAA) in 1992. She directs the NRDC's Coastal Project, which seeks to protect the nation's coasts and monitors the implementation of the Federal Coastal Zone Management Act, the federal offshore oil leasing program, the Oil Pollution Act of 1990, and the Ocean Dumping Ban Act.

Chasis has overseen the preparation of numerous reports, including: *Testing the Waters: A National Perspective on Beach Closings,* which examined beach closings in 22 coastal states, and *Ebb Tide for Pollution,* which clearly and concisely outlined the threats to the ocean and suggested ways individuals can help preserve the sea. She currently oversees NRDC's litigation against Exxon stemming from the *Exxon Valdez* oil spill in Prince William Sound, Alaska, and monitors government efforts to restore the area affected by the spill. She participated in New York Governor Mario Cuomo's Task Force on Coastal Resources, which developed a blueprint for the management of New York State's extensive coastal resources, and has served as chairperson of the Coast

Alliance and on numerous committees of the National Academy of Sciences and the Office of Technology Assessment looking into coastal issues.

Chasis is a 1969 graduate of Smith College and a 1972 graduate of the New York University School of Law. She has been an attorney with the NRDC since 1973.

Barry Commoner

Compared to Rachel Carson in *E Magazine,* Barry Commoner was also called by *Time Magazine* in 1970 "the Paul Revere of Ecology." In 1971, he warned a U.S. Senate Subcommittee on Oceans and the Atmosphere, "The oceans have become the world's sink and the death of the oceans will be the death of us all." Commoner has done more perhaps than any other living scientist to draw public attention to the environmental ills caused by technology. Head of Washington University's Center for the Biology of Natural Systems, Commoner published *Science and Survival* in 1966 and *The Closing Circle* in 1971, in which he explained the relationship between pollution and profits. He also popularized his so-called Laws of Ecology:

> Everything is connected to everything else (the ecosystem consists of multiple interconnected parts that act on one another), everything must go somewhere (there is no such thing as waste), nature knows best (any man-made change in a natural system is likely to be detrimental to that system) and there is no such thing as a free lunch (anything extracted from the global ecosystem by human effort must be replaced or paid for later).

Commoner ran for president on the Citizen's Party ticket in 1980.

Commandant Jacques-Yves Cousteau

This French oceanographer, explorer, inventor, filmmaker, and author, born in 1910, has raised humanity's awareness and knowledge of the oceans since World War II, when he made his first underwater films. In 1946, he was named head of the French Navy's Underwater Research Group. He designed dozens of underwater breathing apparatuses and vehicles, which have permitted the underwater marine research that forms the basis of much of the study of marine pollution. Cousteau spoke eloquently on the dangers of marine pollution as early as 1971 before the

U.S. Senate Subcommittee on Oceans and the Atmosphere. A number of his films, such as *The Silent World,* have won Academy Awards.

His Cousteau Society undertakes dozens of research projects each year, many in the area of marine pollution. Cousteau's contribution to the understanding of the earth's biosphere and humanity's role in preserving the integrity of the planet for future generations is unique. His environmental ethic involves two basic criteria: First, we should not gamble with the survival of the human species or the quality of life of future generations, and second, no chance should be taken on issues that could bring about irreversible damage to the environment. "Without a healthy water system, no life is possible on the planet, and any kind of pollution ends up in the sea."

Jean-Michel Cousteau

Jean-Michel Cousteau, the son of Jacques-Yves Cousteau, was born in 1938. He is no less an advocate for the oceans than is his famous father. During the mid-1960s, through the television series, *The Undersea World of Jacques Cousteau,* Jean-Michel helped teach the world that the ocean has a limited ability to absorb waste before it suffers permanent damage. In 1973, he began an innovative educational field study program called *Project Ocean Search,* which offers people of all ages an opportunity to participate in marine studies.

An architect by profession and training, he has created numerous artificial floating islands, six schools, and a residential/recreational complex in Madagascar. He also designed *Parc Oceanique Cousteau,* a unique public entertainment and educational facility in Paris, where visitors have been treated to exciting underwater experiences. Jean-Michel focuses on public education, complementing and continuing the work of his father: inspiring others to protect and preserve the oceans. He is currently the executive vice president of The Cousteau Society.

Ted Danson

Ted Danson, born in 1947, the star of numerous feature-length films and of the television comedy series "Cheers," founded the American Oceans Campaign (AOC) and actively serves as the organization's president. He first became involved in the field of marine pollution prevention when he was unable to find a clean beach near his home in southern California where his young daughters could swim.

His mission is to preserve and protect marine ecosystems by educating the public and decisionmakers on the need to stop abusing the oceans. His Hollywood renown has helped the cause considerably. He has focused AOC on helping to establish strong public policy. The organization under his direction has been instrumental in the banning of driftnets in American waters. AOC is also active in the coalition that is fighting oil development in the Arctic National Wildlife Refuge, has blocked lease sales for offshore oil and gas development, organized the National Coastal Caucus, worked to end the dumping of inadequately treated sewage and sludge off our coasts, drafted beach protocols, and co-produced conservation videos.

Marjory Stoneman Douglas

Founder and president emeritus of The Friends of the Everglades, Ms. Douglas continues her work to protect this invaluable part of our nation's resources beyond her 100th birthday. Born in Minneapolis, Minnesota, in 1890, Douglas grew up in Massachusetts and graduated from Wellesley College in 1912. She moved to Florida with her father, who founded *The Miami Herald.*

Following years of success as a reporter and freelance writer focused on Florida's unique history and character, Douglas spent the 1940s researching the importance and the beauty of the Florida Everglades, an area nobody had previously considered worth writing about. *The Everglades: River of Grass* was published in 1947. It was considered the textbook for the region for years and played an important role in Douglas's successful, almost single-handed effort to create Everglades National Park. The Everglades is a vital "interface" between the ocean and Florida's freshwater system.

Marjory Douglas is one of America's most honored conservationists, having received honorary doctorates from nine colleges and universities, and numerous awards. In 1989, she was selected as one of *Ms.* magazine's six Women of the Year, placed on *Esquire* magazine's annual list of Women We Love, and designated an ABC News "Person of the Week." The Sierra Club made her an honorary vice president on her 99th birthday. She is currently writing a biography of British naturalist and writer W. H. Hudson.

Sylvia Earle

Marine botanist and environmentalist Sylvia Earle has played an important role in making the ocean depths accessible to scientific study and in assessing the damage caused by oil spills and other environmental disasters. Born in Gibbstown, New Jersey, in 1935, Earle investigated the living creatures in a backyard pond as a child, and took her first dive as a teenager after moving with her family to Clearwater, Florida. She has since spent more than 6,000 hours underwater.

She received a masters degree from Duke University when she was 20 years old and went on to study the biology of algae. After participating in numerous international undersea expeditions, she received her doctorate in 1966. Earle lived in an underwater habitat for two weeks and made a historic "Jim-suit dive" in 1979 to the unprecedented depth of 1,250 feet, where she collected specimens and planted a U.S. flag on the sea floor off the coast of Oahu, Hawaii. She is one of the founders of Deep Ocean Technology, which designs and manufactures ocean submersibles used for scientific research.

In 1989, she played an active role in studying the effects on underwater life of the *Exxon Valdez* oil spill in Alaska's Prince William Sound. The following year she became the first female chief scientist of the National Oceanographic and Atmospheric Administration (NOAA). In this capacity she helped assess the environmental damage caused in the Persian Gulf by Iraq's destruction of Kuwaiti oil wells during the Persian Gulf War. She resigned from this post in February 1992. She has been the recipient of numerous honors, including the Conservation Service Award from the California Academy of Sciences in 1979, the Lowell Thomas Award from the Explorer's Club in 1980, and the David B. Stone Medal from the New England Aquarium in 1989. Several marine plant and animal species have been named after her.

Edward D. Goldberg

Born in Sacramento, California, in 1921, Edward Goldberg received a B.S. degree from the University of California, Berkeley, and a Ph.D. in chemistry from the University of Chicago in 1949. That same year he joined the University of California, San Diego, Scripps Institution of Oceanography. He became known for his study of DDT in the marine environment during the years that followed. He was appointed professor of chemistry in 1961.

In 1970, as a senior NATO Fellow at the Institut Royal des Sciences Naturelles de Belgique in Brussels, Belgium, he investigated pollution of the North Sea. That same year he convened a seminar on *Methods of Detection, Measurement and Monitoring of Pollutants in the Marine Environment* for the United Nations Food and Agriculture Organization (FAO). In 1972, Goldberg directed a workshop sponsored by the National Oceanographic and Atmospheric Administration (NOAA) that resulted in the publication of *Marine Pollution Monitoring: Strategies for a National Program* in 1976. In 1975, he initiated a coastal marine pollution surveillance program called *Mussel Watch* for the U.S. Environmental Protection Agency (EPA). He convened two additional NOAA workshops: *Scientific Problems Relating to Ocean Pollution* and *Assimilative Capacity of U.S. Coastal Waters for Pollutants* and coordinated a workshop for the U.S. National Research Council on *The Disposal of Industrial and Domestic Wastes.*

Dr. Goldberg has served as editor of several journals and monograph series and wrote *A Guide to Marine Pollution* (1972) and *Health of the Oceans* (1976), which provided a starting point for the UNESCO Intergovernmental Oceanographic Commission's Working Committee for the Global Investigation of Pollution in the Marine Environment. He published *Coastal Zone Space* in 1991.

Dr. Goldberg has received numerous awards in the field of marine science over the decades, including the B. H. Ketchum Award in 1984 from the Woods Hole Institution, and the Tyler Prize for environmental achievements in 1989.

Richard S. Golob

An expert in pollution control and emergency response, Richard S. Golob (born 1951) is currently president of World Information Systems and the founder and publisher of *Golob's Oil Pollution Bulletin* and the *Hazardous Materials Intelligence Report.* He also established a comprehensive oil spill incident database that has been used by both industry and government to perform cost-benefit analyses, risk assessment studies, and damage assessment. He has co-edited *The Almanac of Science and Technology: What's New and What's Known* (1990) as well as *The Almanac of Renewable Energy* (1993).

A graduate of Harvard College, Golob was a research coordinator at the Smithsonian Institution Center for Short-Lived Phenomena before founding World Information Systems in 1980. He has completed consulting projects in the areas of oil pollution and hazardous waste management for numerous government agencies and corporations, including IMO/UNEP, for which he analyzed the extent of oil pollution in the Arabian Gulf during the 1980s.

Golob has appeared as an expert witness at U.S. House and Senate subcommittee hearings on environmental issues. Golob and his group were described by *The New York Times* as "the major international coordinator of information on oil spills."

Michael Joseph Herz

This "self-appointed keeper of what's left of San Francisco Bay" was born in 1936 and grew up in Minneapolis, the son of a medical magazine publisher and a bird-watcher. Before he became a "baykeeper," he was a brain researcher and, as such, published numerous books and papers on the effects of toxins on the brains and behavior of various animals. In 1973, he helped found the San Francisco chapter of the Oceanic Society and was named its executive director. He also founded and directed the Water Quality Laboratory and Training Program, a citizen-volunteer monitoring and surveillance program for San Francisco Bay and precursor of Baykeepers.

When Herz became vice president of the National Oceanic Society in 1977, he established several local chapters, led the battle to stop the U.S. Navy from disposing of obsolete nuclear submarines off the California coast, and founded the State of the Bay Conference. In 1983, Mr. Herz became a senior research scientist at the Romberg Center for Environmental Studies (San Francisco State University). When he met John Cronin, the Hudson River keeper, in 1987, he decided the San Francisco Bay needed a similar program, but opted to rely more heavily on citizen volunteers. The Baykeepers program has prompted dozens of prosecutions for Clean Water Act violations.

Thor Heyerdahl

Thor Heyerdahl, the Norwegian anthropologist, explorer, and author, was born in 1914 and is best known for his transoceanic travels on papyrus rafts. His first trip, on the *Kon-Tiki,* took place in 1947. On that 4,100-mile journey across the Pacific (from Peru to Polynesia), he saw no signs of human pollution. Just two decades later, on voyages between Morocco and South America in the Ra I and Ra II, he observed drifting clots of oil floating on the surface on 40 of the 57 days he was at sea. He reported on these findings before the U.S. Senate Subcommittee on the Oceans and the Atmosphere in 1971, as well as in his several books.

How To Kill an Ocean was published in the *Saturday Review.* This eyewitness account of pollution in some of the remotest reaches of our oceans, combined with knowledgeable, thoughtful, even-handed warnings as to the foolhardiness of our continuing pollution of the planet's greatest resource, may have done as much to further the public's awareness of the problem as any effort to date.

Bostwick H. Ketchum

Widely considered the father of the study of marine pollution, Bostwick (Buck) Ketchum (1912–1982) was associated with the Woods Hole Oceanographic Institution for more than 40 years, first as a graduate student, later as a scientist, and from 1962 to 1977 as Associate Director. Like his mentor, the late oceanographer Alfred C. Redfield, Ketchum was equally at home in the physical, chemical, and biological realms.

He began writing papers on ocean dumping in 1948, and his major early contribution was on ocean physical processes as they affect the distribution and dispersion of wastes. He was among the first to use biological criteria in the selection of dump sites, and his broad view of oceanography and appreciation for the interconnectedness of marine phenomena are evident in the more than 70 papers he wrote on subjects ranging from estuarine physics to deep ocean biology. He was active in town government and community affairs, and served as advisor and consultant to numerous federal and international organizations. A lectureship in his honor is awarded annually by the Woods Hole Oceanographic Institution Coastal Research Center.

Albert M. Manville, II

A wildlife biologist by training, Albert M. Manville, II, is currently Director of Science Policy for Defenders of Wildlife, an advocacy organization. He is best known for his work in the area of preventing marine debris and for his leadership role in the ratification of MARPOL-Annex V.

Manville received a B.Sc. degree in Zoology/Ecology in 1969 from Allegheny College, an M.Sc. in Natural Resources and Wildlife management from the University of Wisconsin in 1976, and a Ph.D. in Wildlife Ecology and Management from Michigan State University in 1982. He has taught Wildlife Ecology at the U.S. Department of Agriculture Graduate School since 1983, and was named Staff Wildlife Biologist for Defenders of Wildlife in 1984.

In 1986, Manville reinstituted the Entanglement Network, a coalition of more than 50 national organizations dealing with marine debris and entanglement issues, and he still chairs this important clearinghouse. He also co-founded the Ad Hoc Advisory Committee on Plastics in 1987 to encourage the Navy to stop overboard dumping of plastic.

Dr. Manville has also been instrumental in banning driftnets. He achieved national recognition during the aftermath of the *Exxon Valdez* oil spill. He participated in the drafting of the Marine Plastic Pollution Research and Control Act of 1987, and he testified and lobbied on behalf of this legislation, as well. He also successfully worked for passage of the Driftnets Act (1987), for stronger driftnet language in the Magnuson Act (1990), and for passage of the High Seas Driftnet Fisheries Enforcement Act (1992).

Kathryn Jean O'Hara

Marine biologist Kathryn Jean O'Hara, born in 1960, is today one of the nation's leading environmental experts in the field of marine debris and entanglement. When she joined the Center for Marine Conservation in 1985, she took the concept of cleaning up littered beaches and turned it into a comprehensive international program. The CMC's National Beach Cleanup Campaigns held over the course of the past four years have involved hundreds of thousands of volunteers who not only pick up the man-made solid wastes that have been dumped in the marine environment and washed ashore, but who also catalog the tons of nonbiodegradable garbage collected. This data is then entered in the National Marine Debris Database, founded by Ms. O'Hara and sponsored in part by several federal agencies.

To increase public awareness of marine debris, Ms. O'Hara coordinated the production and distribution of an extensive series of public service advertisements, brochures, posters, slide shows, and videotapes.

She is the author of, among other works, *A Citizen's Guide to Plastics in the Ocean: More Than a Litter Problem,* and she has testified on this topic before the U.S. Congressional Staff, given numerous talks, and chaired many committees on the issue of marine debris. For her outstanding work, she received the 1989 Outstanding Individual Service in Environmental Education Award from the North American Association for Environmental Education.

Claiborne Pell

Known as the father of the Sea Grant College program, Senator Claiborne Pell, born in 1918, has had a long and distinguished political career and has long supported programs beneficial for the sea. Following military service and work with the U.S. State Department, Pell was appointed by President Dwight D. Eisenhower in 1959 as a U.S. delegate to the opening meeting of the Intergovernmental Maritime Consultative Organization (IMCO) in London. This introduction to the dangers of marine pollution may well have encouraged the Rhode Island senator in his energetic sponsorship of a bill to create the Sea Grant College program. Signed into law by President Lyndon Johnson in 1966, this law created a network of colleges and universities that have provided an enormous body of basic research needed for the study of marine pollution.

Pell also co-authored *Challenge of the Seven Seas* (1965) as part of his effort to create the Sea Grant program, and served as chairperson of the subcommittee on ocean space of the U.S. Foreign Affairs Committee. In 1968, he introduced in the Senate a draft treaty intended to clarify legal jurisdiction over the ocean floor. He was possibly the first senator to write a proposed treaty himself. This document influenced the formulation of the Law of the Sea Convention (LOSC, 1982). In it he advocated exploration of underwater seas, the creation of a licensing agency to oversee commercial exploitation of the oceans, the establishment of a sea guard, and a ban on the development of weapons on the ocean floor. Through the decades he has remained a strong champion within the U.S. Senate of the careful exploitation and conservation of the seas, and has received numerous honors for his work in this field.

4

Legal Framework, Facts and Data, and Points of View

THE HEART OF A SERIOUS TREATMENT of a subject may well be documentation; this chapter provides three types of documentation, organized into three sections.

The first section, Legal Framework, starts with a list of key international marine pollution conventions (agreements among nations or by international organizations on behalf of participating nations), that is followed by excerpts from some of the most important of these documents; then a list of major U.S. laws concerning marine pollution, again supplemented with pertinent excerpts.

These agreements and regulations provide the legal framework within which marine pollution has been controlled. Only treaties and legislation dealing directly with marine pollution will be found here; laws that have an indirect impact on marine pollution, such as laws to protect the drinking water supply, are excluded.

The second section, Facts and Data, contains information in the form of tables, figures, and statistics. It is organized around the same six categories of pollution presented in Chapter 1: Sewage, Marine Debris, Toxic Chemicals, Heavy Metals, Oil, and

Radioactive Materials. This section provides a quantitative basis for understanding the scope of marine pollution.

The final section, Points of View, consists of excerpts from key documents and speeches, organized by type of pollution as in the second section, but including two additional categories: What Is Marine Pollution? and Our Changing Perceptions of Marine Pollution. The excerpts are given in approximate chronological order so as to illustrate the evolution of thought regarding marine pollution. This section documents some of the ways in which human treatment of the seas has changed over the decades.

Legal Framework

International Conventions and Agreements

The first known "Law of the Sea" was extremely simple. It was formulated by the seventeenth-century lawyer Grotius. "*Mare Liberum*," he said. "The sea is free." The following century, this absolute freedom of passage was eliminated within territorial waters, defined as a three-mile strip of coastal waters for nations bordering the sea. The intention of this eighteenth-century agreement was simply to allow nations to protect themselves against invaders. The law of the sea remained essentially unchanged for the next 200 years.

During the twentieth century, however, numerous international conventions to protect the oceans have been formulated. The text of five of the most important of these follow Table 1. Although the United Nations Conference on the Human Environment (Sweden, 1972) did not culminate in an international treaty, a number of its key principles concerning the marine environment are also included at the end of this subsection.

TABLE 1 International Marine Pollution Conventions

Convention	Place, Year
Treaty Between His Majesty in Respect of the United Kingdom and the President of the United States of Venezuela Relating to the Submarine Areas of the Gulf of Paria, Caracas	Caracas, 1946
International Convention for the Prevention of Pollution of the Sea by Oil (OILPOL)	Brussels, 1954
Geneva Conventions of 1958 • Convention on the Territorial Sea and Contiguous Zone • Convention on the High Seas • Convention on Fishing and Conservation of the Living Resources of the Sea • Convention on the Continental Shelf	Geneva, 1958
Convention on the Liability of Operators of Nuclear Ships	Brussels, 1962
Treaty Banning Nuclear Weapons Testing in the Atmosphere, in Outer Space and Under Water (Test Ban Treaty)	Moscow, 1963
Resolution 2467 (XXIII)--Committee on the Peaceful uses of the Sea Bed and the Ocean Floor beyond the limits of National Jurisdiction. United Nations	New York, Geneva, 1969
International Convention Relating to Intervention on the High Seas in Cases of Oil Pollution Casualties (Brussels Intervention Convention	Brussels, 1969
Tanker Owners Voluntary Agreement Concerning Liability for Oil Pollution (TOVALOP)	London, 1969
International Convention on Civil Liability for Oil Pollution Damage (CLC)	Brussels, 1969
International Convention on the Establishment of an International Fund for Oil Pollution Damage (Fund Convention)	London, 1971
Contract Regarding an Interim Supplement to Tanker Liability for Oil Pollution (CRISTAL)	London, 1971

Table 1 (*cont.*)

International Convention for the Prevention of Marine Pollution by Dumping of Wastes and Other Matter (Ocean Dumping Convention)	London, Mexico City, Moscow, and Washington, 1972
International Convention for the Prevention of Pollution by Ships (MARPOL)	London, 1973
Protocol on Intervention on the High Seas in Cases of Substances Other Than Oil	London, 1973
Convention for the Prevention of Marine Pollution from Land-based Sources (The Paris Convention)	Paris, 1974
Convention on the Safety of Life at Sea (SOLAS Convention)	London, 1974
Barcelona Convention	Barcelona, 1976
Convention on Civil Liability for Oil Pollution Damage Resulting from Exploration of and Exploitation of Seabed Mine Resources	London, 1977
Kuwait Convention	Kuwait City, 1978
United Nations Convention on the Law of the Sea (LOSC) Includes: • Convention on the High Seas • Convention on the Continental Shelf • Convention on the Territorial Sea and the Contiguous Zone Not ratified	New York, 1982
Convention for the Protection and Development of the Marine Environment of the Wider Caribbean Region (Cartagena Convention)	Cartagena, Colombia, 1983
Second North Sea Conference.	London, 1987
1972 Convention on the Prevention of Marine Pollution by Dumping of Wastes and Other Matter	London, 1990.
Declaration of on the Protection of the Arctic Environment	Finland, 1991

Treaty Excerpts

The first known international treaty that contained provisions for marine pollution prevention was signed in 1942:

Treaty between His Majesty in Respect of the United Kingdom and the President of the United States of Venezuela Relating to the Submarine Areas of the Gulf of Paria, Caracas, February 26, 1942.

Article 7

Each of the High Contracting Parties shall take all practical measures to prevent the exploitation of any submarine areas claimed or occupied by him in the Gulf from causing the pollution of the territorial waters of the other by oil, mud or any other fluid or substance liable to contaminate the navigable waters or the foreshore and shall concert with the other to make said measures as effective as possible.
Entry into force September 22, 1942.

Source: Bernd Rüster and Bruno Simma, eds., *International Protection of the Environment: Treaties and Related Documents,* Vol. 1. Dobbs Ferry, NY: Oceana Publications, Inc., 1975. p. 331.

The first significant, multinational effort to prevent marine pollution was an accord signed in London in 1954 that focused almost exclusively on oil pollution from tankers. Although the International Convention for the Prevention of Pollution of the Sea by Oil was heavily amended over the years and has been superseded by MARPOL (1973), it remains a landmark agreement. In addition to excerpts from the original document of 1954, also included are Resolutions 2 and 12 from the amendments to the Convention made in 1962. These resolutions served to prioritize many of the marine pollution research efforts that took place worldwide over the following years.

International Convention for the Prevention of Pollution of the Sea by Oil, 1954. Done at London, on 12 May 1954.

The Governments represented at the International Conference on Pollution of the Sea by Oil held in London from 26th April 1954, to 12th May 1954,
Desiring to take action by common agreement to prevent pollution of the sea by oil discharged from ships, and considering that this end may best be achieved by the conclusion of a Convention,

Have accordingly appointed the undersigned plenipotentiaries, who, having communicated their full powers, found in good and due form, have agreed as follows:

Article VII

As from a date twelve months after the present Convention comes into force in respect of any of the territories of a Contracting Government all ships registered in that territory shall be required to be so fitted as to prevent the escape of fuel oil or heavy diesel oil into bilges the contents of which are discharged into the sea without being passed through an oily water separator.

Article VIII

As from a date three years after the present Convention comes into force in respect of any of the territories of a Contracting Government, that Government shall ensure the provision in each main port in that territory of facilities adequate for the reception, without causing undue delay to ships, of such residues from oily ballast water and tank washings as would remain for disposal by ships, other than tankers, using the port, if the water had been separated by the use of an oily water separator, a settling tank or otherwise. Each Contracting Government shall from time to time determine which ports are the main ports in its territories for the purposes of this Article, and shall notify the Bureau in writing accordingly indicating whether adequate reception facilities have been installed.

Amendments established in 1962:

Resolution 2
Encouragement of Accessions to the Convention

The International Conference on Prevention of Pollution of the Sea by Oil, 1962,
RECOGNIZING that acceptance and conscientious observance of measures to prevent or control oil pollution by the great majority of ships operating in a given region will be essential in order to effect significant improvement in the oil pollution situation;
RECOGNIZING that prevention of pollution of the seas by oil requires broad international co-operation, including the provision at ports at which ships habitually call of facilities enabling ships to dispose of oily wastes;
BELIEVING that it should be the responsibility of Governments having sea coasts or having sea-going ships under their flag to keep the seas and beaches clean from oil pollution for the enjoyment of the

public and to promote the preservation of the world's wildlife and fish resources,
RESOLVE

(1) that the Contracting Governments to the International Convention for the Prevention of Pollution of the Sea by Oil, 1954, should accept the amendments to the Convention agreed upon by the present Conference at the earliest possible date;

(2) that the Inter-Governmental Maritime Consultative Organization should be asked to call to the attention of its members and of other members of the United Nations or of any of the Specialized Agencies or Parties to the Statute of the International Court of Justice, which are not members of the Organization and which have not become parties to the Convention, the need for their co-operation in the international efforts to that end and invite them to become parties to the Convention;

(3) that to the extent to which it is able the Organization should upon request furnish information and advice to the Governments which have not become parties to the Convention with a view to facilitating their acceptance of the Convention.

Resolution 12

The need for research on prevention of oil pollution
The International Conference on Prevention of Pollution of the Sea by Oil,
Having Noted the information available on the research and development work carried out by many countries,
RESOLVE

(1) That research should continue on many aspects of the prevention of oil pollution of the sea, and specifically on the following subjects:

  (a) oily water separators for use in ships. There does not yet exist a separator, simple and compact enough for use in ships that will deal effectively with any mixtures of persistent oil and water likely to be encountered on board ship including, particularly, mixtures containing oil of specific gravity very close to that of fresh or sea-water;

  (b) devices or measures, other than oily water separators, designed to prevent pollution of the sea by the discharge from ships of persistent oil or oily mixtures;

  (c) methods for the confinement of oil and its removal from the surface of the sea.

Methods based on powders which are used to sink the oil are not viewed with favour on the grounds that such methods are of doubtful practicability and permanence and can cause undesirable fouling of the sea bed. Methods based on emulsifiers have the disadvantage that such agents may be toxic to marine flora and fauna. Certain mechanical methods show great promise for use in calm waters, but are of doubtful value on the open sea;

(d) the development of a device to detect, measure and record the oil content of discharges from ships;

(e) the effects of persistent oils on marine flora and fauna and the role of microorganisms in destroying such oils;

(2) that the results of research on the above and allied subjects (including but not limited to technical information concerning research and experiments, including shipboard research with respect to anti-pollution measures or devices) should be transmitted annually by the Government concerned to the Inter-Governmental Maritime Consultative Organization for collation and transmission to all Contracting Governments, and that technical matters involving research should be referred to the technical experts of the Contracting Governments.

Source: Bernd Rüster and Bruno Simma, eds., *International Protection of the Environment: Treaties and Related Documents,* Vol. 1. Dobbs Ferry, NY: Oceana Publications, Inc., 1975. pp. 332–377.

The next significant international treaty, commonly known as the Ocean Dumping Convention, was negotiated in 1972. It was far more specific than any previous agreement regarding the substances considered harmful to the marine environment, as illustrated in Article 12 and the Annexes included below.

Convention on the Prevention of Marine Pollution by Dumping of Wastes and Other Matter. London, Mexico City, Moscow, and Washington, December 29, 1972.

The Contracting Parties to this Convention,
Recognizing that the marine environment and the living organisms which it supports are of vital importance to humanity, and all people have an interest in assuring that it is so managed that its quality and resources are not impaired;

Recognizing that the capacity of the sea to assimilate wastes and render them harmless, and its ability to regenerate natural resources, is not unlimited;
Recognizing that States have, in accordance with the Charter of the United Nations and the principles of international law, the sovereign right to exploit their own resources pursuant to their own environmental policies, and the responsibility to ensure that activities within their jurisdiction or control do not cause damage to the environment of other States or of areas beyond the limits of national jurisdiction;
Recalling resolution 2749 (XXV) of the General Assembly of the United Nations on the principles governing the sea-bed and the ocean floor and the subsoil thereof, beyond the limits of national jurisdiction;
Noting that marine pollution originates in many sources, such as dumping and discharges through the atmosphere, rivers, estuaries, outfalls and pipelines, and that it is important that States use the best practicable means to prevent such pollution and develop products and processes which will reduce the amount of harmful wastes to be disposed of;
Being convinced that international action to control the pollution of the sea by dumping can and must be taken without delay but that this action should not preclude discussion of measures to control other sources of marine pollution as soon as possible; and
Wishing to improve protection of the marine environment by encouraging States with a common interest in particular geographical areas to enter into appropriate agreements supplementary to this Convention;
Have agreed as follows:

Article I

Contracting Parties shall individually and collectively promote the effective control of all sources of pollution of the marine environment, and pledge themselves especially to take all practicable steps to prevent the pollution of the sea by the dumping of waste and other matter that is liable to create hazards to human health, to harm living resources and marine life, to damage amenities or to interfere with other legitimate uses of the sea.

Article II

Contracting parties shall, as provided for in the following Articles, take effective measures individually, according to their scientific, technical and economic capabilities, and collectively, to prevent marine pollution caused by dumping and shall harmonize their policies in this regard.

Article IV

1. In accordance with the provisions of this Convention, Contracting Parties shall prohibit the dumping of any wastes or other matter in whatever form or condition except as otherwise specified below:

   (a) The dumping of wastes or other matter listed in Annex I is prohibited:

   (b) The dumping of wastes or other matter listed in Annex II requires a prior special permit;

   (c) The dumping of all other wastes or matter requires a prior general permit.

2. Any permit shall be issued only after careful consideration of all the factors set forth in Annex III, including prior studies of the characteristics of the dumping site as set forth in Sections B and C of that Annex.

3. No provision of this Convention is to be interpreted as preventing a Contracting Party from prohibiting, insofar as that Party is concerned, the dumping of wastes or other matter not mentioned in Annex I. That Party shall notify such measures to the Organization.

Article XII

The Contracting Parties pledge themselves to promote, within the competent specialised agencies and other international bodies, measures to protect the marine environment against pollution caused by:

(a) Hydrocarbons including oil, and their wastes;

(b) Other noxious or hazardous matter transported by vessels for purposes other than dumping;

(c) Wastes generated in the course of operation of vessels, aircraft, platforms and other man-made structures at sea;

(d) Radio-active pollutants from all sources, including vessels;

(e) Agents of chemical and biological warfare;

(f) Wastes or other matter directly arising from, or related to the exploration, exploitation and associated off-shore processing of sea-bed mineral resources.

The Parties will also promote, within the appropriate international organisation, the codification of signals to be used by vessels engaged in dumping.

Annex I

1. Organohalogen compounds.
2. Mercury and mercury compounds.
3. Cadmium and cadmium compounds.
4. Persistent plastics and other persistent synthetic materials, for example, netting and ropes, which may float or may remain in suspension in the sea in such a manner as to interfere materially with fishing, navigation or other legitimate uses of the sea.
5. Crude oil, fuel oil, heavy diesel oil, and lubricating oils, hydraulic fluids, and any mixtures containing any of these, taken on board for the purpose of dumping.
6. High-level radio-active wastes or other high-level radio-active matter, defined on public health, biological or other grounds, by the competent international body in this field, at present the International Atomic Energy Agency, as unsuitable for dumping.
7. Materials in whatever form (e. g. solids, liquids, semi-liquids, gases or in a living state) produced for biological and chemical warfare.
8. The preceding paragraphs of this Annex do not apply to substances which are rapidly rendered harmless by physical, chemical or biological processes in the sea provided they do not:

    (i) Make edible marine organisms unpalatable, or

    (ii) Endanger human health or that of domestic animals. The consultative procedure provided for under Article XIV should be followed by a Party if there is doubt about the harmlessness of the substance.

9. This Annex does not apply to wastes or other materials (e. g. sewage sludges and dredged spoils) containing the matters referred to in paragraphs 1–5 above as trace contaminants. Such wastes shall be subject to the provisions of Annexes II and III as appropriate.

Annex II

The following substances and materials requiring special care are listed for the purposes of Article VI I (a).

A. Wastes containing significant amounts of the matters listed below:

    Arsenic

    Lead

    Copper

Zinc

Organosilicon compounds

Cyanides

Fluorides

Pesticides and their by products not covered in Annex I

(and their compounds)

B. In the issue of permits for the dumping of large quantities of acids and alkalis, consideration shall be given to the possible presence in such wastes of the substances listed in paragraph A and to the following additional substances:

Beryllium

Chromium

Nickel

Vanadium

(and their compounds)

C. Containers, scrap metal and other bulky wastes liable to sink to the sea bottom which may present a serious obstacle to fishing or navigation.

D. Radio-active wastes or other radio-active matter not included in Annex I. In the issue of permits for the dumping of this matter, the Contracting Parties should take full account of the recommendations of the competent international body in this field, at present the International Atomic Energy Agency.

Source: Bernd Rüster and Bruno Simma, eds., *International Protection of the Environment: Treaties and Related Documents,* Vol. 2. Dobbs Ferry, NY: Oceana Publications, Inc., 1991. pp. 537–544.

The International Convention for the Prevention of Pollution from Ships (MARPOL, 1973) superseded the 1954 International Convention for the Prevention of Pollution of the Sea by Oil. It took into consideration a broad range of potential pollutants, those which originate at sea as well as those from land-based sources. Its provisions remain in effect today. For Annexes I through IV, titles only are given here, but significant portions of Annex V, which forms the basis for the marine debris portion of U.S. Public Law 100–220, are reproduced below.

INTERNATIONAL CONVENTION FOR THE PREVENTION OF POLLUTION FROM SHIPS, 1973 (MARPOL)

THE PARTIES TO THE CONVENTION,
BEING CONSCIOUS of the need to preserve the human environment in general and the marine environment in particular,
RECOGNIZING that deliberate, negligent or accidental release of oil and other harmful substances from ships constitutes a serious source of pollution,
RECOGNIZING ALSO the importance of the International Convention for the Prevention of Pollution of the Sea by Oil, 1954, as being the first multilateral instrument to be concluded with the prime objective of protecting the environment, and appreciating the significant contribution which that Convention has made in preserving the seas and coastal environment from pollution,
DESIRING to achieve the complete elimination of intentional pollution of the marine environment by oil and other harmful substances and the minimization of accidental discharge of such substances,
CONSIDERING that this object may best be achieved by establishing rules not limited to oil pollution having a universal purport,
HAVE AGREED as follows:

Article 6

Detection of Violations and Enforcement of the Convention

(1) Parties to the Convention shall co-operate in the detection of violations and the enforcement of the provisions of the present Convention, using all appropriate and practicable measures of detection and environmental monitoring, adequate procedures for reporting and accumulation of evidence.

(2) A ship to which the present Convention applies may, in any port or off-shore terminal of a Party, be subject to inspection by officers appointed or authorized by that Party for the purpose of verifying whether the ship has discharged any harmful substances in violation of the provisions of the Regulations. If an inspection indicates a violation of the Convention, a report shall be forwarded to the Administration for any appropriate action.

(3) Any Party shall furnish to the Administration evidence, if any, that the ship has discharged harmful substances or effluents containing such substances in violation of the provisions of the Regulations. If it is practicable to do so, the competent authority of the former Party shall notify the Master of the ship of the alleged violation.

(4) Upon receiving such evidence, the Administration so informed shall investigate the matter, and may request the other Party to

furnish further or better evidence of the alleged contravention. If the Administration is satisfied that sufficient evidence is available to enable proceedings to be brought in respect of the alleged violation, it shall cause such proceedings to be taken in accordance with its law as soon as possible. The Administration shall promptly inform the Party which has reported the alleged violation, as well as the Organization, of the action taken.

(5) A Party may also inspect a ship to which the present Convention applies when it enters the ports or off-shore terminals under its jurisdiction, if a request for an investigation is received from any Party together with sufficient evidence that the ship has discharged harmful substances or effluents containing such substances in any place. The report of such investigation shall be sent to the Party requesting it and to the Administration so that the appropriate action may be taken under the present Convention.

Article 8
Reports on Incidents Involving Harmful Substances

(1) A report of an incident shall be made without delay to the fullest extent possible in accordance with the provisions of Protocol I to the present Convention.

(2) Each Party to the Convention shall:

(a) make all arrangements necessary for an appropriate officer or agency to receive and process all reports on incidents; and

(b) notify the Organization with complete details of such arrangements for circulation to other Parties and member States of the Organization.

(3) Whenever a Party receives a report under the provisions of the present Article that Party shall relay the report without delay to:

(a) the Administration of the ship involved; and

(b) any other State which may be affected.

(4) Each Party to the Convention undertakes to issue instructions to its maritime inspection vessels and aircraft and to other appropriate services, to report to its authorities any incident referred to in Protocol I to the present Convention. That Party shall, if it considers it appropriate, report accordingly to the Organization and to any other party concerned.

Annex I: Regulations for the Prevention of Pollution by Oil

Annex II: Regulations for the Control of Pollution by Noxious Liquid Substances in Bulk

Annex III: Regulations for the Prevention of Pollution by Harmful Substances Carried by Sea in Packaged Forms, or in Freight Containers, Portable Tanks or Road and Rail Tank Wagons

Annex IV: Regulations for the Prevention of Pollution by Sewage From Ships

Annex V
REGULATIONS FOR THE PREVENTION OF POLLUTION BY GARBAGE FROM SHIPS

Regulation 1
Definitions

For the purposes of this Annex:

(1) "Garbage" means all kinds of victual, domestic and operational waste excluding fresh fish and parts thereof, generated during the normal operation of the ship and liable to be disposed of continuously or periodically except those substances which are defined or listed in other Annexes to the present Convention . . .

(3) "Special area" means a sea area where for recognized technical reasons in relation to its oceanographical and ecological condition and to the particular character of its traffic the adoption of special mandatory methods for the prevention of sea pollution by garbage is required. Special areas shall include those listed in Regulation 5 of this Annex.

Regulation 2
Application

The provisions of this Annex shall apply to all ships.

Regulation 3
Disposal of Garbage outside Special Areas

(1) Subject to the provisions of Regulations 4, 5 and 6 of this Annex:

(a) the disposal into the sea of all plastics, including but not limited to synthetic ropes, synthetic fishing nets and plastic garbage bags is prohibited;

(b) the disposal into the sea of the following garbage shall be made as far as practicable from the nearest land but in any case is prohibited if the distance from the nearest land is less than:

(i) 25 nautical miles for dunnage, lining and packing materials which will float;

(ii) 12 nautical miles for food wastes and all other garbage including paper products, rags, glass, metal, bottles, crockery and similar refuse;

(c) disposal into the sea of garbage specified in sub-paragraph (2) of this Regulation may be permitted when it has passed through a comminuter or grinder and made as far as practicable from the nearest land but in any case is prohibited if the distance from the nearest land is less than 3 nautical miles. Such comminuted or ground garbage shall be capable of passing through a screen with openings no greater than 25 millimeters.

(2) When the garbage is mixed with other discharges having different disposal or discharge requirements the more stringent requirements shall apply . . .

Regulation 4
Special Requirements for Disposal of Garbage

(1) Subject to the provisions of paragraph (2) of this Regulation, the disposal of any materials regulated by this Annex is prohibited from fixed or floating platforms engaged in the exploration, exploitation and associated offshore processing of seabed mineral resources, and from all other ships when alongside or within 500 metres of such platforms.

(2) The disposal into the sea of food wastes may be permitted when they have been passed through a comminuter or grinder from such fixed or floating platforms located more than 12 nautical miles from land and all other ships when alongside or within 500 metres of such platforms. Such comminuted or ground food wastes shall be capable of passing through a screen with openings no greater than 25 millimeters.

Regulation 5
Disposal of Garbage Within Special Areas

(1) For the purposes of this Annex the special areas are the Mediterranean Sea area, the Baltic Sea area, the Black Sea area, the Red Sea area and the "Gulfs area" . . .

(2) Subject to the provisions of Regulation 6 of this Annex:

(a) disposal into the sea of the following is prohibited:

(i) all plastics, including but not limited to synthetic ropes, synthetic fishing nets and plastic garbage bags; and

(ii) all other garbage, including paper products, rags, glass, metal, bottles, crockery, dunnage, lining and packing materials;

(b) disposal into the sea of food wastes shall be made as far as practicable from land, but in any case not less than 12 nautical miles from the nearest land.

(3) When the garbage is mixed with other discharges having different disposal or discharge requirements the more stringent requirements shall apply.

(4) Reception facilities within special areas:

(a) The Government of each Party to the Convention, the coastline of which borders a special area undertakes to ensure that as soon as possible in all ports within a special area, adequate reception facilities are provided in accordance with Regulation 7 of this Annex, taking into account the special needs of ships operating in these areas.

(b) The Government of each Party concerned shall notify the Organization of the measures taken pursuant to sub-paragraph (a) of this Regulation. Upon receipt of sufficient notifications the Organization shall establish a date from which the requirements of this Regulation in respect of the area in question shall take effect. The Organization shall notify all Parties of the date so established no less than twelve months in advance of that date.

(c) After the date so established, ships calling also at ports in these special areas where such facilities are not yet available, shall fully comply with the requirements of this Regulation.

Regulation 6
Exceptions

Regulations 3, 4 and 5 of this Annex shall not apply to:

(a) the disposal of garbage from a ship necessary for the purpose of securing the safety of a ship and those on board or saving life at sea; or

(b) the escape of garbage resulting from damage to a ship or its equipment provided all reasonable precautions have been taken before and after the occurrence of the damage, for the purpose of preventing or minimizing the escape; or

(c) the accidental loss of synthetic fishing nets or synthetic material incidental to the repair of such nets, provided that all reasonable precautions have been taken to prevent such loss.

Regulation 7
Reception Facilities

(1) The Government of each Party to the Convention undertakes to ensure the provision of facilities at ports and terminals for the

reception of garbage, without causing undue delay to ships, and according to the needs of the ships using them.

(2) The Government of each Party shall notify the Organization for transmission to the Parties concerned of all cases where the facilities provided under this Regulation are alleged to be inadequate.

Source: Bernd Rüster and Bruno Simma, eds., *International Protection of the Environment: Treaties and Related Documents,* Vol. 2. Dobbs Ferry, NY: Oceana Publications, Inc., 1991. pp. 552–657.

The United Nations Convention on the Law of the Sea (LOSC) represented a complete overhaul of the international legal protection of the world's oceans and was the result of an arduous fifteen-year process. Although it was finalized in Jamaica in 1982, it has still not been ratified by the United States and a number of other nations. Disagreements center around seabed mineral exploration and extraction. Part XII of the LOSC, the section most relevant to marine pollution issues, is reproduced below:

The United Nations Convention on the Law of the Sea Part XII. Protection and Preservation of the Marine Environment

SECTION 1. GENERAL PROVISIONS

Article 192
General Obligation

States have the obligation to protect and preserve the marine environment.

Article 193
Sovereign Right of States to Exploit Their Natural Resources

States have the sovereign right to exploit their natural resources pursuant to their environmental policies and in accordance with their duty to protect and preserve the marine environment.

Article 194
Measures to Prevent, Reduce and Control Pollution of the Marine Environment

1. States shall take, individually or jointly as appropriate, all measures consistent with this Convention that are necessary to prevent,

reduce and control pollution of the marine environment from any source, using for this purpose the best practicable means at their disposal and in accordance with their capabilities, and they shall endeavor to harmonize their policies in this connection.

2. States shall take all measures necessary to ensure that activities under their jurisdiction or control are so conducted as not to cause damage by pollution to other States and their environment, and that pollution arising from incidents or activities under their jurisdiction or control does not spread beyond the areas where they exercise sovereign rights in accordance with this Convention.

3. The measures taken pursuant to this Part shall deal with all sources of pollution of the marine environment. These measures shall include, *inter alia,* those designed to minimize to the fullest possible extent:

   (a) the release of toxic, harmful or noxious substances, especially those which are persistent, from land-based sources, from or through the atmosphere or by dumping;

   (b) pollution from vessels, in particular measures for preventing accidents and dealing with emergencies, ensuring the safety of operations at sea, preventing, intentional and unintentional discharges, and regulating the design, construction, equipment, operation and manning of vessels;

   (c) pollution from installations and devices used in exploration or exploitation of the natural resources of the sea-bed and subsoil, in particular measures for preventing accidents and dealing with emergencies, ensuring the safety of operations at sea, and regulating the design, construction, equipment, operation and manning of such installations or devices;

   (d) pollution from other installations and devices operating in the marine environment, in particular measures for preventing accidents and dealing with emergencies, ensuring the safety of operations at sea, and regulating the design, construction, equipment, operation and manning of such installations or devices.

4. In taking measures to prevent, reduce or control pollution of the marine environment, States shall refrain from unjustifiable interference with activities carried out by other States in the exercise of their rights and in pursuance of their duties in conformity with this Convention.

5. The measures taken in accordance with this Part shall include those necessary to protect and preserve rare or fragile ecosystems as well as the habitat of depleted, threatened or endangered species and other forms of marine life.

Article 195
Duty Not To Transfer Damage or Hazards or Transform One Type of Pollution into Another

In taking measures to prevent, reduce and control pollution of the marine environment, States shall act so as not to transfer, directly or indirectly, damage or hazards from one area to another or transform one type of pollution into another.

Article 196
Use of Technologies or Introduction of Alien or New Species

1. States shall take all measures necessary to prevent, reduce and control pollution of the marine environment resulting from the use of technologies under their jurisdiction or control, or the intentional or accidental introduction of species, alien or new, to a particular part of the marine environment, which may cause significant and harmful changes thereto.
2. This article does not affect the application of this Convention regarding the prevention, reduction and control of pollution of the marine environment.

SECTION 2. GLOBAL AND REGIONAL COOPERATION

Article 197
Co-operation on a Global or Regional Basis

States shall co-operate on a global basis and, as appropriate, on a regional basis, directly or through competent international organizations, in formulating and elaborating international rules, standards and recommended practices and procedures consistent with this Convention, for the protection and preservation of the marine environment, taking into account characteristic regional features.

Article 198
Notification of Imminent or Actual Damage

When a State becomes aware of cases in which the marine environment is in imminent danger of being damaged or has been damaged by pollution, it shall immediately notify other States it deems likely to be affected by such damage, as well as the competent international organizations.

Article 199
Contingency Plans Against Pollution

In the cases referred to in article 198, States in the area affected, in accordance with their capabilities, and the competent international organizations shall co-operate, to the extent possible, in eliminating the effects of pollution and preventing or minimizing the damage. To this end, States shall jointly develop and promote contingency plans for responding to pollution incidents in the marine environment.

Article 200
Studies, Research Programmes and Exchange of Information and Data

States shall co-operate, directly or through competent international organizations, for the purpose of promoting studies, undertaking programmes of scientific research and encouraging the exchange of information and data acquired about pollution of the marine environment. They shall endeavour to participate actively in regional and global programmes to acquire knowledge for the assessment of the nature and extent of pollution, exposure to it, and its pathways, risks and remedies.

Article 201
Scientific Criteria for Regulations

In the light of the information and data acquired pursuant to article 200, States shall co-operate, directly or through competent international organizations, in establishing appropriate scientific criteria for the formulation and elaboration of rules, standards and recommended practices and procedures for the prevention, reduction and control of pollution of the marine environment.

SECTION 3. TECHNICAL ASSISTANCE

Article 202
Scientific and Technical Assistance to Developing States

States shall, directly or through competent international organizations:

(a) promote programmes of scientific, educational, technical and other assistance to developing States for the protection and preservation of the marine environment and the prevention, reduction and control of marine pollution. Such assistance shall include, *inter alia*:

(i) training of their scientific and technical personnel;

(ii) facilitating their participation in relevant international programmes;

(iii) supplying them with necessary equipment and facilities;

(iv) enhancing their capacity to manufacture such equipment;

(v) advice on and developing facilities for research, monitoring, educational and other programmes;

(b) provide appropriate assistance, especially to developing States, for the minimization of the effects of major incidents which may cause serious pollution of the marine environment;

(c) provide appropriate assistance, especially to developing States, concerning the preparation of environmental assessments.

Article 203
Preferential Treatment for Developing States

Developing States shall, for the purposes of prevention, reduction and control of pollution of the marine environment or minimization of its effects, be granted preference by international organizations in:

(a) the allocation of appropriate funds and technical assistance; and

(b) the utilization of their specialized services.

SECTION 4. MONITORING AND ENVIRONMENTAL ASSESSMENT

Article 204
Monitoring of the Risks or Effects of Pollution

1. States shall, consistent with the rights of other States, endeavour, as far as practicable directly or through the competent international organizations, to observe, measure, evaluate and analyse, by recognized scientific methods, the risks or effects of pollution on the marine environment.
2. In particular, States shall keep under surveillance the effects of any activities which they permit or in which they engage in order to determine whether these activities are likely to pollute the marine environment.

Article 205
Publication of Reports

States shall publish reports of the results obtained pursuant to Article 204 or provide such reports at appropriate intervals to the competent international organizations, which should make them available to all States.

Article 206
Assessment of Potential Effects of Activities

When States have reasonable grounds for believing that planned activities under their jurisdiction or control may cause substantial pollution of or significant and harmful changes to the marine environment, they shall, as far as practicable, assess the potential effects of such activities on the marine environment and shall communicate reports of the results of such assessments in the manner provided in Article 205.

SECTION 5. INTERNATIONAL RULES AND NATIONAL LEGISLATION TO PREVENT, REDUCE AND CONTROL POLLUTION OF THE MARINE ENVIRONMENT

Article 207
Pollution From Land-based Sources

1. States shall adopt laws and regulations to prevent, reduce and control pollution of the marine environment from land-based sources, including rivers, estuaries, pipelines and outfall structures, taking into account internationally agreed rules, standards and recommended practices and procedures.
2. States shall take other measures as may be necessary to prevent, reduce and control such pollution.
3. States shall endeavour to harmonize their policies in this connection at the appropriate regional level.
4. States, acting especially through competent international organizations or diplomatic conference, shall endeavour to establish global and regional rules, standards and recommended practices and procedures to prevent, reduce and control pollution of the marine environment from land-based sources, taking into account characteristic regional features, the economic capacity of developing States and their need for economic development. Such rules, standards and recommended practices and procedures shall be reexamined from time to time as necessary.
5. Laws, regulations, measures, rules, standards and recommended practices and procedures referred to in paragraphs 1, 2 and 4 shall include those designed to minimize, to the fullest extent possible, the release of toxic, harmful or noxious substances, especially those which are persistent, into the marine environment.

. . . .

Article 221
Measures To Avoid Pollution Arising From Maritime Casualties

1. Nothing in this Part shall prejudice the right of States, pursuant to international law, both customary and conventional, to take and enforce measures beyond the territorial sea proportionate to the actual or threatened damage to protect their coastline or related interests, including fishing, from pollution or threat of pollution following upon a maritime casualty or acts relating to such a casualty, which may reasonably be expected to result in major harmful consequences.
2. For the purposes of this article, maritime casualty means a collision of vessels, stranding or other incident of navigation, or other occurrence on board a vessel or external to it resulting in material damage or imminent threat of material damage to a vessel or cargo.

Article 222
Enforcement with Respect to Pollution From or Through the Atmosphere

States shall enforce, within the air space under their sovereignty or with regard to vessels flying their flag or vessels or aircraft of their registry, their laws and regulations adopted in accordance with article 212, paragraph 1, and with other provisions of this Convention and shall adopt laws and regulations and take other measures necessary to implement applicable international rules and standards established through competent international organizations or diplomatic conference to prevent, reduce and control pollution of the marine environment from or through the atmosphere, in conformity with all relevant international rules and standards concerning the safety of air navigation.

. . . .

Article 236
Sovereign Immunity

The provisions of this Convention regarding the protection and preservation of the marine environment do not apply to any warship, naval auxiliary, other vessels or aircraft owned or operated by a State and used, for the time being, only on government noncommercial service. However, each State shall ensure, by the adoption of appropriate measures not impairing operations or operational capabilities of such vessels or aircraft owned or operated by it, that such vessels or aircraft act in a manner consistent, so far as is reasonable and practicable, with this Convention.

Source: Bernd Rüster and Bruno Simma, eds., *International Protection of the Environment: Treaties and Related Documents,* Vol. 2 (II. Marine Pollution A.1. International Agreements and Related Documents II/A/10–12–82). Dobbs Ferry, NY: Oceana Publications, Inc., 1991. pp. 9–27.

The United Nations Conference on the Human Environment (UNCHE) was held in Sweden in 1972. This landmark environmental conference signaled the coming-of-age of the environmental movement, the critical juncture at which environmentalism became international in scope and a basic concern of governments and leaders around the world.

Declaration of the United Nations Conference on the Human Environment

The United Nations Conference on the Human Environment,
Having met at Stockholm from 5 to 16 June 1972,

Having considered the need for a common outlook and for common principles to inspire and guide the peoples of the world in the preservation and enhancement of the human environment,
Proclaims that:

1. Man is both creature and monster of his environment, which gives him physical sustenance and affords him the opportunity for intellectual, moral, social and spiritual growth. In the long and tenuous evolution of the human race on this planet a stage has been reached when, through the rapid acceleration of science and technology, man has acquired the power to transform his environment in countless ways and on an unprecedented scale. Both aspects of man's environment, the natural and the man-made, are essential to his well-being and to the enjoyment of basic human rights—even the right to life itself.

2. The protection and improvement of the human environment is a major issue which affects the well-being of peoples and economic development throughout the world; it is the urgent desire of the peoples of the whole world and the duty of all Governments . . .

6. A point has been reached in history when we must shape our actions throughout the world with a more prudent care for their environmental consequences. Through ignorance or indifference we can do massive and irreversible harm to the earthly environment on which our life and well-being depend. Conversely, through fuller knowledge and wiser action, we can achieve for ourselves and our posterity a better life in an environment more in keeping with human needs and hopes . . . A growing class of environmental problems, because they are regional or global in extent or because they affect the common international realm, will require extensive co-operation among nations and action by international organizations in the common interest. The Conference calls upon Governments and peoples to exert common efforts for the preservation and improvement of the human environment, for the benefit of all the people and for their posterity.

Principles

States the common conviction that:
Principle 1
Man has the fundamental right to freedom, equality and adequate conditions of life, in an environment of quality that permits a life of dignity and well-being, and he bears a solemn responsibility to protect and improve the environment for present and future generations . . .
Principle 2
The natural resources of the earth, including the air, water, land, flora

and fauna and especially representative samples of natural ecosystems, must be safeguarded for the benefit of present and future generations through careful planning or management, as appropriate.

Principle 3

The capacity of the earth to produce vital renewable resources must be maintained and, wherever practicable, restored or improved.

Principle 4

Man has a special responsibility to safeguard and wisely manage the heritage of wildlife and its habitat which are now gravely imperiled by a combination of adverse factors. Nature conservation, including wildlife must therefore receive importance in planning for economic development.

Principle 5

The non-renewable resources of the Earth must be employed in such a way as to guard against the danger of their future exhaustion and to ensure that benefit from such employment are shared by all mankind.

Principle 6

The discharge of toxic substances or of other substances and the release of heat, in such quantities or concentrations as to exceed the capacity of the environment to render them harmless, must be halted in order to ensure that serious or irreversible damage is not inflicted upon ecosystems. The just struggle of the peoples of all countries against pollution should be supported.

Principle 7

States shall take all possible steps to prevent pollution of the seas by substances that are liable to create hazards to human health, to harm living resources and marine life, to damage amenities or to interfere with other legitimate uses of the sea.

Source: *Report of the United Nations Conference on the Human Environment*. Stockholm, June 5–16, 1972, New York 1973. pp. 3–7.

U.S. Marine Pollution Laws

An Act of Congress in 1888 made it illegal to dump construction rubbish into New York Harbor. This was the first U.S. legislation to attempt to prevent marine pollution. After much debate and delay, the Rivers and Harbors Act of 1899 extended this regulation to all navigable waterways in the United States. Two decades passed before any further legislation was approved, and as Table 2 reveals, significant marine pollution prevention efforts at the legislative level in the United States did not begin in earnest until the 1960s. "P.L." preceding the number along the right-hand column of Table 2 stands for "Public Law." Excerpts from five key pieces of U.S. legislation complete this subsection.

Table 2 U.S. Marine Pollution Legislation

Act of Congress makes it illegal to dump construction rubbish in New York Harbor	1886
Rivers and Harbors Act of 1899	1899
Oil Pollution Act, 1924	Pub. No. 238, 1924
Oil Pollution Act of 1961	P.L. 87-167, 1961
Clean Waters Restoration Act of 1966	P.L.89-753, 1966
Marine Resources and Engineering Development Act of 1966,	P.L. 89-454, 1966
National Environmental Policy Act of 1969 (NEPA)	P.L. 91-190, 1969
Federal Water Pollution Control Act Amendments of 1972 (Clean Water Act)	P.L. 92-500, 1972
Coastal Zone Management Act of 1972	P.L. 92-583, 1972
Marine Mammal Protection Act of 1972	P.L. 92-522, 1972
Ports and Waterways Safety Act of 1972	P.L. 92-340, 1972
Marine Protection, Research, and Sanctuaries Act (Ocean Dumping Act)	P.L. 92-532, 1972
Deepwater Port Act of 1974	P.L. 93-627
Intervention on the High Seas Act	P.L. 93-248, 1974
Toxic Substances Control Act	P.L. 94-469, 1976
Fishery Conservation and Management Act of 1976	P.L. 94-265, 1976
National Ocean Pollution Research and Development and Monitoring Planning Act of 1978	P.L. 95-273, 1978

Table 2 (*cont.*)

Marine Plastic Pollution Control Act of 1987 (MPPRCA). Same as Annex V MARPOL	P.L. 100-220 (Title II), 1987
Marine Science, Technology, and Policy Development Act of 1987	P.L. 100-220 (Title III), 1987
Driftnet Impact Monitoring, Assessment, and Control Act of 1987	P.L. 100-220 (Title IV),1987
Organotin Antifouling Pint Control Act of 1988	P.L. 100-333, 1988
Ocean Dumping Ban Act of 1988	P.L. 100-688 (Title I) 1988
United States Public Vessel Medical Waste Anti-Dumping Act of 1988	P.L.100-688 (Title III) 1988
Shore Protection Act of 1988	P.L.100-688 (Title IV) 1988
Degradable Plastic Ring Carriers	P.L. 100-556 (Title I) 1988
The Oil Pollution Act of 1990	P.L. 101-380

The Control of Certain Pollutants in New York Harbor, 1888

The placing, discharging, or despoiling, by any process or in any manner, of refuse, dirt, ashes, cinders, mud, sand, dredgings, sludge, acid, or any other matter of any kind, other than that flowing from streets, sewers, and passing therefrom in a liquid state, in the tidal waters of the harbor of New York, or its adjacent or tributary waters, or in those of Long Island Sound, within the limits which shall be prescribed by the supervisor of the harbor, is strictly forbidden, and every such act is made a misdemeanor, and every person engaged in or who shall aid, abet, authorize, or instigate a violation of this section, shall upon conviction, be punishable by fine or imprisonment, or both, such fine to be not less than $250 nor more than $2,500, and the imprisonment to be not less than thirty days nor more than one year, either or both united, as the judge before whom conviction is obtained shall decide, one-half of said fine to be paid to the person or persons giving information which shall lead to conviction of this misdemeanor.

Source: Act of June 28, 1888, in *Statutes at Large,* ch. 496, sec. 1, 25, p. 209.

The Control of Certain Pollutants in Navigable Waters was the first federal marine pollution control legislation approved by the U.S. Congress that covered more than a single body of water. The bill's main focus was allocation of funds for waterway improvement. It is only Section 13, excerpted in full below, that attempts to establish pollution controls. Note that it does not prohibit raw sewage from being dumped directly into any waterway:

The Control of Certain Pollutants in Navigable Waters, 1899

Sec. 13. It shall not be lawful to throw, discharge, or deposit, or cause, suffer, or procure to be thrown, discharged, or deposited either from or out of any ship, barge, or other floating craft of any kind, or from the shore, wharf, manufacturing establishment, or mill of any kind, any refuse matter of any kind or description whatever other than that flowing from streets and sewers and passing therefrom in a liquid state, into any navigable water of the United States, or into any tributary of any navigable water from which the same shall float or be washed into such navigable water; and it shall not be lawful to deposit, or cause, suffer, or procure to be deposited material of any kind in any place on the bank of any navigable water, or on the bank of any tributary of any navigable water, where the same shall be liable to be washed into such navigable water, either by ordinary or high tides, or by storms or floods, or otherwise, whereby navigation shall or may be impeded or obstructed: *Provided,* That nothing herein contained shall extend to, apply to, or prohibit the operations in connection with the improvement of navigable waters or construction of public works, considered necessary and proper by the United States officers supervising such improvement or public work: *And provided further,* That the Secretary of the Army, whenever in the judgment of the Chief of Engineers anchorage and navigation will not be injured thereby, may permit the deposit of any material above mentioned in navigable waters, within limits to be defined and under conditions to be prescribed by him, provided application is made to him prior to depositing such material and whenever any permit is so granted the conditions thereof shall be strictly complied with and any violation thereof shall be unlawful.

Source: Act of March 3, 1899, in *Statutes at Large*, ch. 425, sec. 13, p. 1152.

The Ocean Dumping Act of 1972 prohibits the dumping of a broad range of pollutants in U.S. territorial waters, including radioactive substances. The research provisions under Title II have guided the Sea Grant program, as well as those of many other federally-mandated ocean research efforts.

The Ocean Dumping Act of 1972

TITLE I—OCEAN DUMPING

Prohibited Acts

Sec. 101. (a) No person shall transport from the United States any radiological, chemical or biological warfare agent or any high-level radioactive waste, or except as may be authorized in a permit issued under this title, and subject to regulations issued under section 108 hereof by the Secretary of the Department in which the Coast Guard is operating, any other material for the purpose of dumping it into ocean waters.

(b) No person shall dump any radiological, chemical, or biological warfare agent or high-level radioactive waste, or except as may be authorized in a permit issued under this title, any other material transported from any location outside of the United States (l) into the territorial sea of the United States, or (2) into a zone contiguous to the territorial sea of the United States, extending to a line twelve nautical miles seaward from the base line from which the breadth of the territorial sea is measured, to the extent that it may affect the territorial sea or the territory of the United States.

(c) No officer, employee, agent, department, agency, or instrumentality of the United States shall transport from any location outside the United States any radiological, chemical, or biological warfare agent or high-level radioactive waste, or, except as may be authorized in a permit issued under this title, any other material for the purpose of dumping it into ocean waters.

TITLE II—COMPREHENSIVE RESEARCH ON OCEAN DUMPING

Sec. 201. The Secretary of Commerce, in coordination with the Secretary of the Department in which the Coast Guard is operating and with the Administrator shall, within six months of the enactment of this Act, initiate a comprehensive and continuing program of monitoring and research regarding the effects of the dumping of material into ocean waters or other coastal waters where the tide ebbs and flows or into the Great Lakes or their connecting waters and shall report from time to time, not less frequently than annually, his findings (including an evaluation of the short-term ecological effects and the social and economic factors involved) to the Congress.

Sec. 202. (a) The Secretary of Commerce, in consultation with other appropriate Federal departments, agencies, and instrumentalities shall, within six months of the enactment of this Act, initiate a comprehensive and continuing program of research with respect to the possible long-range effects of pollution, overfishing, and man-induced changes of ocean ecosystems. In carrying out such research, the Secretary of Commerce shall take into account such factors as existing and proposed international policies affecting oceanic problems, economic considerations involved in both the protection and the use of the oceans, possible alternatives to existing programs, and ways in which the health of the oceans may best be preserved for the benefit of succeeding generations of mankind.

(b) In carrying out his responsibilities under this section, the Secretary of Commerce, under the foreign policy guidance of the President and pursuant to international agreements and treaties made by the President with the advice and consent of the Senate, may act alone or in conjunction with any other nation or group of nations and shall make known the results of his activities by such channels of communication as may appear appropriate.

(c) In January of each year, the Secretary of Commerce shall report to the Congress on the results of activities undertaken by him pursuant to this section during the previous fiscal year.

(d) Each department, agency, and independent instrumentality of the Federal Government is authorized and directed to cooperate with the Secretary of Commerce in carrying out the purposes of this section and, to the extent permitted by law, to furnish such information as may be requested.

(e) The Secretary of Commerce, in carrying out his responsibilities under this section, shall, to the extent feasible utilize the personnel, services and facilities of other Federal departments, agencies, and instrumentalities (including those of the Coast Guard for monitoring purposes), and is authorized to enter into appropriate inter-agency agreements to accomplish this action.

Sec. 203. The Secretary of Commerce shall conduct and encourage, cooperate with, and render financial and other assistance to appropriate public (whether Federal, State, interstate, or local) authorities, agencies, and institutions, private agencies and institutions, and individuals in the conduct of, and to promote the coordination of, research, investigations, experiments, training, demonstrations, surveys and studies for the purpose of determining means of minimizing or ending all dumping of materials within five years of the effective date of this Act.

Sec. 204. There are authorized to be appropriated for the first fiscal year after this Act is enacted and for the next two fiscal years thereafter such sums as may be necessary to carry out this title, but the sums appropriated for any such fiscal year may not exceed $6,000,000.

TITLE III—MARINE SANCTUARIES

Sec. 301. Not withstanding the provisions of subsection (h) of section 3 of this Act, the term "Secretary," when used in this title, means Secretary of Commerce.

Sec. 302. (a) The Secretary, after consultation with the Secretaries of State, Defense, the Interior, and Transportation, the Administrator, and the heads of other interested Federal agencies, and with the approval of the President, may designate as marine sanctuaries those areas of the ocean waters, as far seaward as the outer edge of the Continental Shelf, as defined in the Convention of the Continental Shelf (15 U.S.T. 74; TIAS 5578), of other coastal waters where the tide ebbs and flows, or of the Great Lakes and their connecting waters, which he determines necessary for the purpose of preserving or restoring such areas for their conservation, recreational, ecological, or esthetic values. The consultation shall include an opportunity to review and comment on a specific proposed designation.

Source: *United States Statutes at Large*. Washington, DC: United States Government Printing Office. Vol. 86, p. 1060.

Public Law 100–220 is an enormous piece of legislation. It encompasses no fewer than five broad areas of concern, grouped under "Titles." Title I has to do with U.S.–Japan fishing accords, Title II provides the legislative basis of the Marine Plastic Pollution Control Act (MPPCA), Title III provides further direction for the research efforts of the Sea Grant program, Title IV is the first driftnet legislation enacted in the United States, and Title V focuses on the red tide problems that plagued areas along the Atlantic shoreline. Only the introduction and outline of this act is reproduced below, to demonstrate its breadth and unprecedented scope.

Public Law 100–220—December 29, 1987

An Act
To provide congressional approval of the Governing International Fishery Agreement between the United States and Japan; to implement the provisions of Annex V to the International Convention for the Prevention of Pollution from Ships, 1973, to reauthorize the National Sea Grant College Program Act, to improve efforts to monitor, assess, and reduce the adverse impacts of driftnet, and for other purposes.
Be it enacted by the Senate and House of Representatives of the United States of America in Congress assembled,

Sec. 1. Short Title

This Act may be cited as the "United States–Japan Fishery Agreement Approval Act of 1987."

Sec. 2. Table of Contents

The contents of this Act are as follows:

Sec. 1. Short title.

Sec. 2. Table of contents.

Title I—Approval of Governing International Fishery Agreement with Japan

Sec. 1001. Approval of agreement.

Title II—Plastic Pollution Research and Control

Sec. 2001. Short title.

Sec, 2002. Effective date.

Sec. 2003. Preemption; additional State requirements.

Subtitle A—Amendments to Act to Prevent Pollution From Ships

Sec. 2101. Definitions.

Sec. 2102. Application of Act.

Sec. 2103. Pollution reception facilities.

Sec. 2104. Violations.

Sec. 2105. Civil penalties.

Sec. 2106. Proposed amendments to protocol.

Sec. 2107. Administration and enforcement; refuse record books; waste management plans; notification of crew and passengers.

Sec. 2108. Compliance with international law.

Subtitle B—Studies and Report

Sec. 2201. Compliance report.

Sec. 2202. EPA study of methods to reduce plastic pollution.

Sec. 2203. Effects of plastic materials on the marine environment.

Sec. 2204. Plastic pollution public education program.

Subtitle C—New York Bight

Sec. 2301. New York Bight restoration plan.

Sec. 2302. New York Bight plastic study.

Sec. 2303. Reports.

Sec. 2304. Definitions.

Sec. 2305. Authorization of appropriations.

Title III—Marine Science, Tecnology, and Policy Development

Sec. 3001. Short title.

Subtitle A—National Sea Grant College Program Authorization

Sec. 3101. Short title.

Sec. 3102. Reference to the National Sea Grant College Program Act.

Sec. 3103. Declaration of policy.

Sec. 3104. Definitions.

Sec. 3105. Contracts and grants.

Sec. 3106. Sea grant strategic research program.

Sec. 3107. Fellowships.

Sec. 3108. Sea grant review panel.

Sec. 3109. Marine affairs and resource management improvement grants.

Sec. 3110. Authorization of appropriations.

Sec. 3111. Sea grant international program.

Subtitle B—Great Lakes Mapping

Sec. 3201. Short title.

Sec. 3202. Great Lakes shoreline mapping plan.

Sec. 3203. Preparation of Great Lake shoreline maps.

Sec. 3204. Contract authority.

Sec. 3205. Definitions.

Sec. 3206. Authorization of appropriations.

Title IV—Driftnet Impact Monitoring, Assessment, and Control

Sec. 4001. Short title.

Sec. 4002. Findings.

Sec. 4003. Definitions.

Sec. 4004. Monitoring agreements.

Sec. 4005. Impact report.

Sec. 4006. Enforcement agreements.

Sec. 4007. Evaluations and recommendations.

Sec. 4008. Construction with other laws.

Sec. 4009. Authorization of appropriations.

Title V—Red Tide Contamination

Sec. 5001. Declaration of disaster.

Sec. 5002. Provision of assistance.

Sec. 5003. Recent North Carolina Coast red tide contamination, defined.

Source: *United States Statutes at Large*. Washington, DC: United States Government Printing Office. Vol. 101, p. 1459.

The Oil Pollution Act of 1990 was a direct response to the *Exxon Valdez* oil spill in Prince William Sound, Alaska. It attempts to establish conditions that will help prevent further large spills in pristine areas and to establish penalties that will provide economic incentives for industry to meet and surpass these conditions. The table of contents of this extensive piece of legislation is reproduced below:

Public Law 101–380—August 18, 1990

An Act
To establish limitations on liability for damages resulting from oil pollution, to establish a fund for the payment of compensation for such damages, and for other purposes.
Be it enacted by the Senate and House of Representatives of the United States of America in Congress assembled,

Sec. 1. Short Title

This Act may be cited as the "Oil Pollution Act of 1990."

Sec. 2. Table of Contents

The contents of this Act are as follows:

Title I—Oil Pollution Liability and Compensation

Sec. 1001. Definitions.

Sec. 1002. Elements of liability.

Sec. 1003. Defenses to liability.

Sec. 1004. Limits on liability.

Sec. 1005. Interest.

Sec. 1006. Natural resources.

Sec. 1007. Recovery by foreign claimants.

Sec. 1008. Recovery by responsible party.

Sec. 1009. Contribution.

Sec. 1010. Indemnification agreements.

Sec. 1011. Consultation on removal actions.

Sec. 1012. Uses of the Fund.

Sec. 1013. Claims procedure.

Sec. 1014. Designation of source and advertisement.

Sec. 1015. Subrogation.

Sec. 1016. Financial responsibility.

Sec. 1017. Litigation, jurisdiction, and venue.

Sec. 1018. Relationship to other law.

Sec. 1019. State financial responsibility.

Sec. 1020. Application.

Title II—Conforming Amendments

Sec. 2001. Intervention on the High Seas Act.

Sec. 2002. Federal Water Pollution Control Act.

Sec. 2003. Deepwater Port Act.

Sec. 2004. Outer Continental Shelf Lands Act Amendments of 1978.

Title III—International Oil Pollution Prevention and Removal

Sec. 3001. Sense of Congress regarding participation in international regime.

Sec. 3002. United States–Canada Great Lakes oil spill cooperation.

Sec. 3003. United States–Canada Lake Champlain oil spill cooperation.

Sec. 3004. International inventory of removal equipment and personnel.

Sec. 3005. Negotiations with Canada concerning tug escorts in Puget Sound.

Title IV—Prevention and Removal

Subtitle A—Prevention

Sec. 4101. Review of alcohol and drug abuse and other matters in issuing licenses, certificates of registry, and merchant mariners' documents.

Sec. 4102. Term of licenses, certificates of registry, and merchant mariners' documents, criminal record reviews in renewals.

Sec. 4103. Suspension and revocation of licenses, certificates of registry, and merchant mariners' documents for alcohol and drug abuse.

Sec. 4104. Removal of master or individual in charge.

Sec. 4105. Access to National Driver Register.

Sec. 4106. Manning standards for foreign tank vessels.

Sec. 4107. Vessel traffic service systems.

Sec. 4108. Great Lakes pilotage.

Sec. 4109. Periodic gauging of plating thickness of commercial vessels.

Sec. 4110. Overfill and tank level or pressure monitoring devices.

Sec. 4111. Study on tanker navigation safety standards.

Sec. 4112. Dredge modification study.

Sec. 4113. Use of liners.

Sec. 4114. Tank vessel manning.

Sec. 4115. Establishment of double hull requirement for tank vessels.

Sec. 4116. Pilotage.

Sec. 4117. Maritime pollution prevention training program study.

Sec. 4118. Vessel communication equipment regulations.

Subtitle B—Removal

Sec. 4201. Federal removal authority.

Sec. 4202. National planning and response system.

Sec. 4203. Coast Guard vessel design.

Sec. 4204. Determination of harmful quantities of oil and hazardous substances.

Sec. 4205. Coastwise oil spill response endorsements.

Subtitle C—Penalties and Miscellaneous

Sec. 4301. Federal Water Pollution Control Act penalties.

Sec. 4302. Other penalties.

Sec. 4303. Financial responsibility civil penalties.

Sec. 4304. Deposit of certain penalties into oil spill liability trust fund.

Sec. 4305. Inspection and entry.

Sec. 4306. Civil enforcement under Federal Water Pollution Control Act.

Title V—Prince Willlam Sound Provisions

Sec 5001. Oil spill recovery institute.

Sec 5002. Terminal and tanker oversight and monitoring.

Sec. 5003. Bligh Reef light.

Sec. 5004. Vessel traffic service system.

Sec. 5005. Equipment and personnel requirements under tank vessel and facility response plans.

Sec. 5006. Funding.

Sec. 5007. Limitation.

Title VI—Miscellaneous

Sec. 6001. Savings provisions.

Sec. 6002. Annual appropriations.

Sec. 6001. Outer Banks protection.

Sec. 6004. Cooperative development of common hydrocarbon-bearing areas.

Title VII—Oil Pollution Research and Development Program

Sec. 7001. Oil pollution research and development program.

Title VIII—Trans-Alaska Pipeline System

Sec. 8001. Short title.

Subtitle A—Improvements to Trans-Alaska Pipeline System

Sec. 8101. Liability within the State of Alaska and cleanup efforts.

Sec. 8102. Trans-Alaska Pipeline Liability Fund.

Sec. 8103. Presidential task force.

Subtitle B—Penalties

Sec. 8201. Authority of the Secretary of the Interior to impose penalties on Outer Continental Shelf facilities.

Sec. 8202. Trans-Alaska pipeline system civil penalties.

Subtitle C—Provisions Applicable to Alaska Natives

Sec. 8301. Land conveyances.

Sec. 8302. Impact of potential spills in the Arctic Ocean on Alaska Natives.

Title IX—Amendments to Oil Spill Liability Trust Fund, etc.

Sec. 9001. Amendments to Oil Spill Liability Trust Fund.

Sec. 9002. Changes relating to other funds.

Source: *United States Statutes at Large*. Washington, DC: U.S. Government Printing Office. Vol. 104, p. 104.

Facts, Statistics, Tables, and Figures

General Facts and Statistics

A fifth of the world's population lives in the world's ten most populous cities; Tokyo, New York, London, Moscow, Shanghai, Bombay, Sao Paulo, Rio de Janeiro, and Peking. All of them except Moscow are situated on the coast, as are most other large cities. A United States government report forecasts that 50 percent of the U.S. population will be living within a hundred miles of the seashore and the Great Lakes by the year 2000.

Source: Colin Moorcraft, *Must the Seas Die?* Boston: Gambit, 1973. p. 7.

Seventy percent of the world's oxygen is renewed through the biological action of minute phytoplankton near the ocean's surface—where 90 percent of its organic life is found.

Source: Jonathan Bartlett, *The Ocean Environment.* New York: The H.W. Wilson Co., 1977. p. 135.

Ninety percent of all marine species are concentrated above the continental shelves next to land. The water above these littoral shelves represents an area of only 8 percent of the total ocean surface, which itself represents only 4 percent of the total body of water, and means that much less than half a percent of the ocean space represents the home of 90 percent of all marine life. This concentration of marine life in shallow waters next to the coasts happens to coincide with the area of concentrated dumping and the outlet of all sewers and polluted river mouths, not to mention silt from chemically treated farmland.

Source: Thor Heyerdahl, "How To Kill an Ocean," in *Saturday Review* (Nov. 29) 1975. pp. 12–18.

Over 80 percent of all ocean pollution comes from land-based activities. This pollution reaches the oceans either from "point" or "nonpoint" sources. Point sources include pipes, ditches, canals, or similar channels that regularly release pollution in a specific area. Sewage and industrial waste are commonly introduced into waters from point sources. Nonpoint pollution covers all types of unregulated runoff from the land, including runoff from urban and agricultural areas.

Source: Walter H. Corson, ed. (The Global Tomorrow Coalition). *The Global Ecology Handbook: What You Can Do about the Environmental Crisis.* Boston: Beacon Press, 1990. p. 138.

Here's a likely scenario of what may happen over the next 40 years. Less developed countries will be responsible for more than 90 percent of the world's new population. . . . Calculations show that these developments—population growth, access to polluting technologies, and higher consumption levels—will interact to raise global waste generation from approximately 2.5 billion metric tons in 1985 to more than 4.5 billion tons by 2025. Whereas poorer countries contributed only 25 percent to global waste in 1985, they will have likely contributed to more than 50 percent by 2025!

Source: R. Paul Shaw, "Population Growth: Is It Ruining the Environment?" in *Populi,* June 1989 (Vol. 16, No. 2), pp. 20–29.

Tables and Figures

Figure 1 reveals ships to be the primary source of marine pollution, although land facilities/pipeline and production/storage facility pollution of coastal waters began to increase sharply at the end of the 1970s.

A single type of marine pollutant can enter the ocean in a variety of ways, from a number of different sources. For example, the nitrogen compounds in acid rain have the same effect on the

FIGURE 1 All Pollution Discharge Volumes in and around U.S. Coastal Waters by Major Source

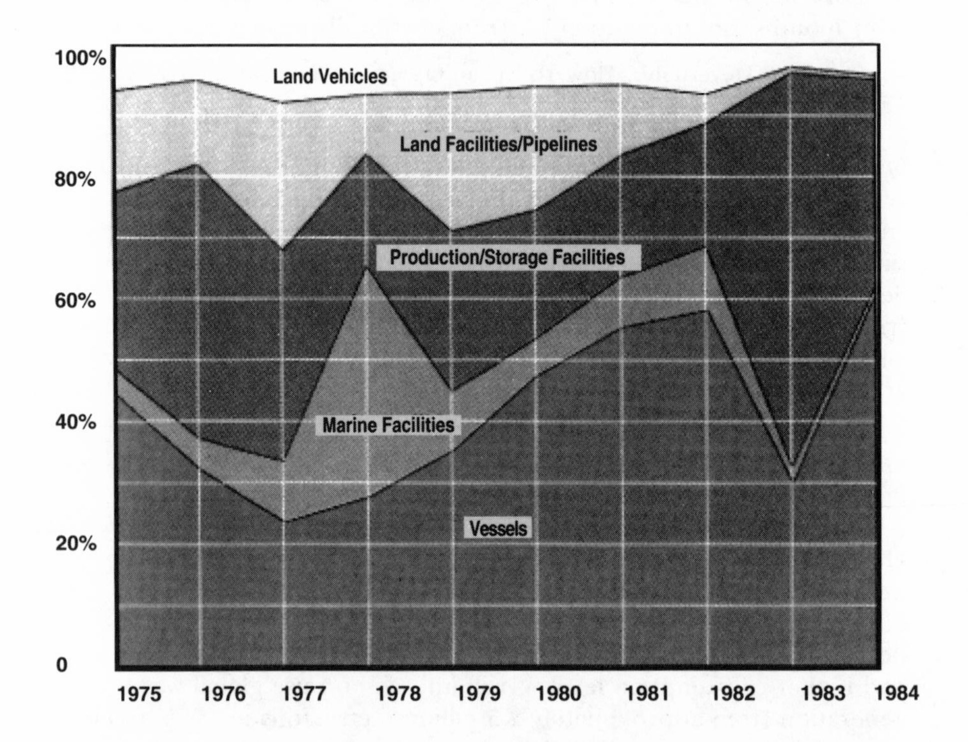

Source: Department of Transportation and Coast Guard. *Poluting Incidents in and around U.S. Waters: Calendar Year 1983 and 1984.* Springfield, VA: National Technical Information Services, 1986, pp. 7, 11, 21.

marine environment as fertilizers and sewage. Industrial waste can contain toxic chemicals, heavy metals, and oil products, producing more than one effect on the ocean and its flora and fauna. Table 3 summarizes the multiple sources and effects of six categories of marine pollutants.

Sewage

Facts and Statistics

. . . Each year, 2.3 trillion gallons of sewage pours into America's coastal waters; and population growth continues to add increased stress to

TABLE 3 Major Pollutants Affecting U.S. Coastal Waters

Pollutant	Source	Impact
Nutrients, inc. nitrogen compounds	Fertilizer, sewags, acid rain from motor vehicles and power plants	Creates algal blooms, destroys marine life
Chlorinated hydrocarbons: pesticides, DDT, PCBs	Agricultural runoff, industrial waste	Contaminates and harms fish and shellfish
Petroleum hydrocarbons	Oil spills, industrial discharge, urban runoff	Kills or harms marine life, damages ecosystems
Heavy Metals: arsenic,cadmium, copper, lead, zinc	Industrial waste, mining	Contaminates and harms fish
Soil and other particulate matter	Soil erosion from construction and farming; dredging, dying algae	Smothers shellfish beds, blocks light needed by marine plant life
Plastics	Ship dumping, household waste, litter	Strangles, mutilates wildlife, damages natural habitats

Source: Walter H. Corson, ed. (The Global Tomorrow Coalition). *The Global Ecology Handbook: What You Can Do about the Environmental Crisis.* Boston: Beacon Press, 1990. p. 142. Reprinted by permission of Beacon Press.

our coastal communities. Nationally, by the year 2010, the number of individuals living in coastal areas is expected to increase from 80 million today to 127 million. This accounts for an estimated 60 percent increase in coastal population.

Source: John F. Kerry, "Omnibus Budget Reconciliation Act—Conference Report," *Congressional Record* (27 October 1990). Daily ed. pH17493-17540.

Treatment plants discharged 3.3 trillion gallons of sewage into marine waters in 1980. That volume is projected to rise to 5.4 trillion gallons by the year 2000.

Source: Beth Millemann, *And Two If by Sea: Fighting the Attack on America's Coasts*. Washington, DC: Coast Alliance, 1986. p. 28.

Because much of the U.S. coast is densely populated, about 35 percent of all U.S. sewage ends up in marine waters.

Source: Walter H. Corson, ed. (The Global Tomorrow Coalition). *The Global Ecology Handbook: What You Can Do about the Environmental Crisis*. Boston: Beacon Press, 1990. p. 138.

Tables and Figures

It is not possible for humans to stop producing biological waste. The way in which human waste is treated, and the ways in which it might be utilized, or recycled, are thus extremely important. Primary, secondary, and tertiary treatment all produce sewage sludge, as illustrated in Figure 2.

FIGURE 2 The Waste Process

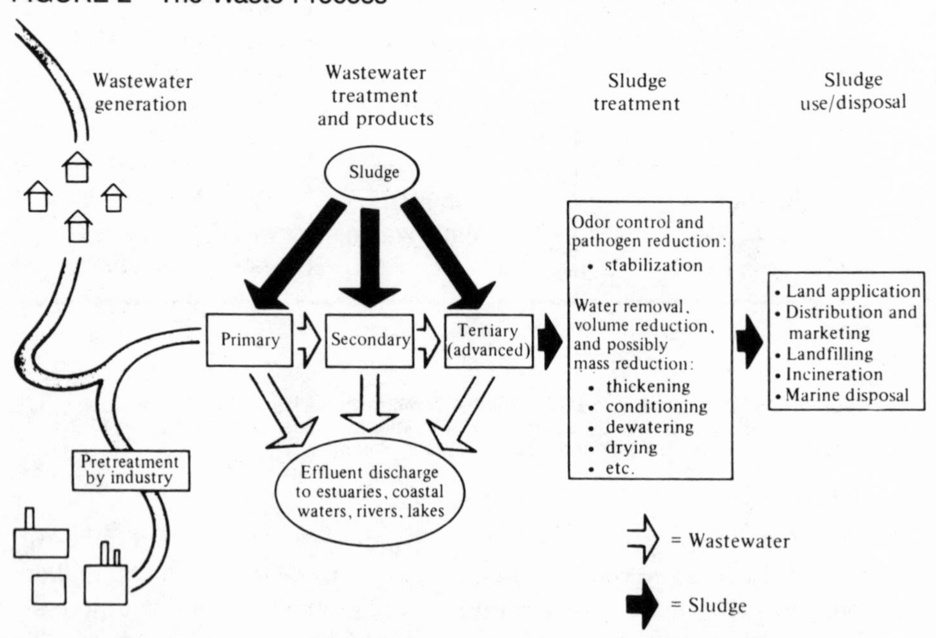

Source: Adapted from U.S. Environmental Protection Agency. Inter-Agency Sludge Task Force. *Use and Disposal of Municipal Sludge*. Washington, DC: EPA, September, 1984 by Gerard J. Mangone in *Marine Policy for America* (2nd Edition), NY: Taylor & Francis, 1988. p. 256. Reproduced with permission.

Significant improvements in sewage treatment took place during the 1970s and 1980s. Table 4 documents the number of people served by primary treatment, and by secondary and/or tertiary treatment plants in 22 developed and semideveloped nations around the world. Sweden, Denmark, and New Zealand have the highest percentage of citizens served by sewage treatment facilities.

TABLE 4 Population Served by Wastewater Treatment Plants in Selected Countries, 1970 to 1985

	Primary Only				Primary + Secondary and/or Tertiary				Total to be Served			
	1970	**1975**	**1980**	**1985**	**1970**	**1975**	**1980**	**1985**	**1970**	**1975**	**1980**	**1985**
Canada	--	--	13.0	10.0	--	--	43.0	47.0	--	49.0	56.0	57.0
USA	--	23.0	17.0	15.0	--	44.0	53.0	59.0	42.0	67 0	70.0	74.0
Japan	--	--	--	--	16.0	23.0	30.0	36.0	16.0	23.0	30.0	36.0
New Zealand	--	9.0	10.0	8.0	--	47.0	49.0	80.0	52.0	56.0	59.0	88.0
Denmark	31.9	29.0	--	20.0	22.4	41.6	--	70.0	54.3	70.6	--	90.0
Finland	5.0	3.0	0.0	0.0	22.0	47.0	65.0	69.0	27.0	50.0	65.0	69.0
Germany	20.5	18.4	10.2	7.5	41.3	56.4	71.6	79.0	61.8	74.8	81.8	86.5
Luxembourg	23.0	--	16.0	14.0	5.0	--	65.0	69.0	28.0	--	81.0	83.0
Netherlands	--	8.0	7.0	4.0	--	37.0	61.0	77.0	--	45.0	68.0	81.0
Norway	1.0	2.0	2.0	3.0	21.0	25.0	37.0	48.0	22.0	27.0	39.0	51.0
Portugal	1.0	2.0	3.0	3.5	2.1	4.0	7.0	8.5	3.1	6.0	10.0	12.0
Spain	--	7.0	8.8	13.2	--	7.3	9.1	15.8	--	14.3	17.9	29.0
Sweden	12.0	4.0	1.0	1.0	66.0	94.0	98.0	98.0	78.0	98.0	99.0	99.0
Switzerland	--	--	--	--	35.0	55.0	70.0	81.0	35.0	55.0	70.0	81.0
UK	--	--	6.0	6.0	--	--	76.0	77.0	--	--	82.0	83.0

Source: Excerpted from a table in the *Environmental Data Compendium 1987,* based on data provided by the Organisation for Economic Cooperation.

Figure 3 shows an almost continuous increase in the amount of sewage sludge dumped in U.S. ocean waters from 1973 to 1982. Sewage sludge is the residue of waste processing facilities. Because industrial waste often goes down the same sewers as human waste, sewage sludge usually contains high levels of toxic chemicals and heavy metals. International treaties and legislation passed in the United States during the 1970s barring the direct ocean

FIGURE 3 Sewage Sludge Dumped in U.S. Coastal Waters from 1973 to 1986

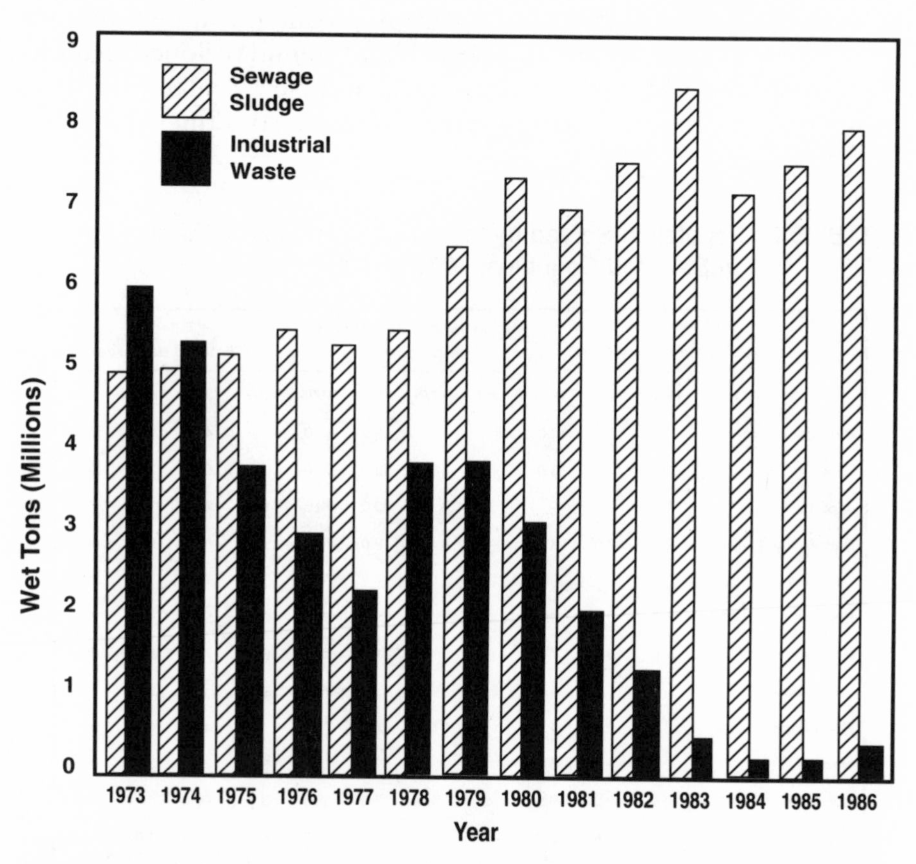

Source: Frits Van der Leeden, Fred L. Troise, and David Keith Todd, eds., *The Water Encyclopedia.* Chelsea, MI: Lewis Publishers, Inc., 1990. p. 605 (adapted from the U.S. Environmental Protection Agency).

dumping of industrial wastes have caused a significant decline in the amount of industrial wastes entering U.S. ocean waters over this same period.

Figure 4 documents U.S. marine disposal in coastal waters surrounding the United States during the mid-1980s. There are fewer facilities discharging wastes into northeastern coastal area waters than anywhere else in the nation, and the Gulf of Mexico coastline has the greatest number of dischargers. Note that this figure refers only to the number of facilities, and does not indicate the amount discharged.

FIGURE 4 Point Source Discharges to U.S. Coastal Waters

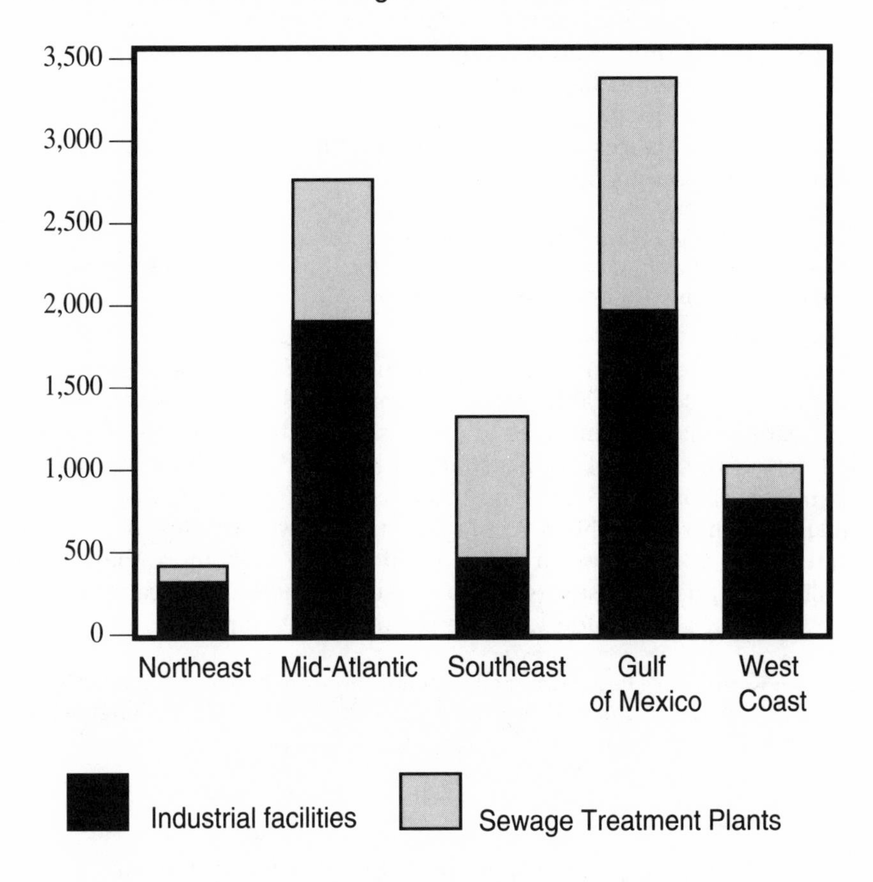

Source: Council on the Environment. *Environmental Quality, 22nd Annual Report.* Washington, DC: GPO, March 1992. Page 34, (from the National Oceanographic and Atmospheric Administration (NOAA). *Estuaries of the United States.* Washington, DC: NOAA, 1990).

Marine Debris

Facts and Statistics

By 1960, more than 6 billion pounds of plastic were produced annually just in the U.S. By 1988, that figure had risen tenfold, to nearly 60 billion pounds—more than 10 pounds of plastic for every person on earth. In 1987, more than 200 million pounds of plastic was used in the U.S. for disposable diapers alone.

Source: Kathryn O'Hara, Suzanne Iudicello, and Rose Bierce, *A Citizen's Guide to Plastics in the Ocean: More Than a Litter Problem.* 3rd rev. ed. Washington, DC: Center for Marine Conservation, 1988. p. 5.

A sea turtle was found with 15 plastic bags in its stomach, and a whale ate no fewer than 50. A turtle found in New York had swallowed 590 yards of fishing line. And a young hawksbill sea turtle found in Hawaii had ingested not just bags and line, but also a plastic flower, part of a bottle cap, a comb, polystyrene chips (probably from a dock), dozens of small, round pieces of plastic, and a plastic golf tee. Of his total weight of 12 pounds, nearly 2 were plastic.

Source: Kathryn O'Hara, Suzanne Iudicello, and Rose Bierce, *A Citizen's Guide to Plastics in the Ocean: More Than a Litter Problem.* 3rd rev. ed. Washington, DC: Center for Marine Conservation, 1988. pp. 21–22.

As many as 50,000 northern fur seal pups have died in a single year due to entanglement, although this number is dropping as the total fur seal population continues to diminish. An abandoned 9-mile-long gill net off the coast of Alaska snagged hundreds of salmon and more than 300 sea birds before being retrieved by environmentalists. It is likely that more than 30,000 king crab traps have been lost in Alaskan waters over the past two or three decades; each year, more than 1.5 million pounds of lobster (worth at least $2.5 million) are caught in these sturdy, almost indestructible plastic "ghost traps."

Source: Kathryn O'Hara, Suzanne Iudicello, and Rose Bierce, *A Citizen's Guide to Plastics in the Ocean: More Than a Litter Problem.* 3rd rev. ed. Washington, DC: Center for Marine Conservation, 1988. p. 16.

Tables and Figures

During beach cleanup efforts in 1989, 540 tons of plastic debris were collected from 3,000 miles of shoreline in the United States, Canada, and Mexico. Figure 5 shows that the contribution by volume of plastic (63 percent of all beach litter) is more than five times as great as that of the next largest component of beach litter: glass.

Figure 6 graphically explains the garbage disposal limitations imposed by the provisions of MARPOL Annex V.

Toxic Chemicals

Facts and Statistics

Each year, factories dump an estimated 5 trillion gallons of waste water into our coastal waters . . .

Source: John F. Kerry, "Omnibus Budget Reconciliation Act—Conference Report," *Congressional Record* (27 October 1990). Daily ed. pH17493-17540.

FIGURE 5 Beach Waste by Type

Percent Composition of Debris Reported during the 1992 International Coastal Cleanups

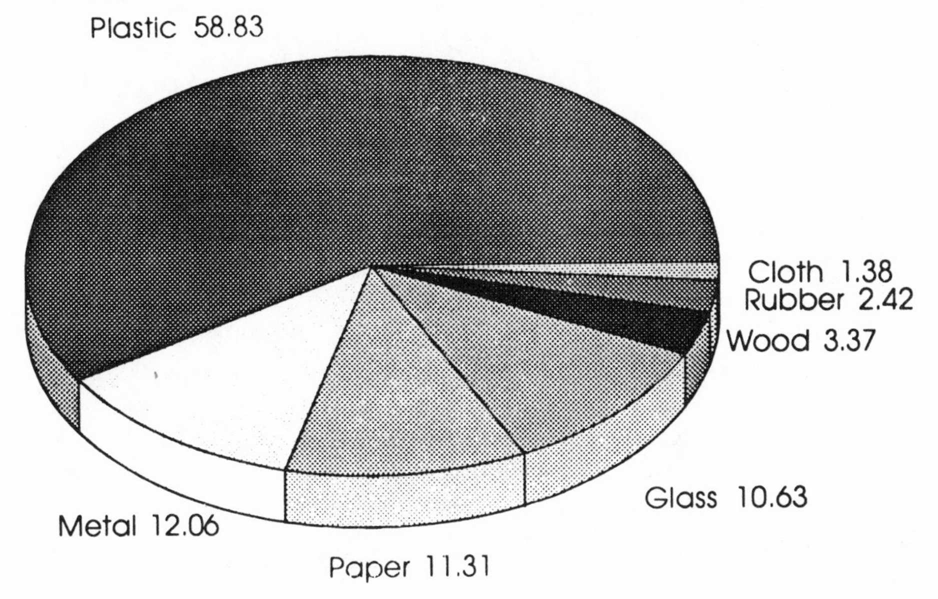

Source: *Garbage,* September/October 1990, p. 26, from the Center for Marine Conservation.

FIGURE 6 Marpol Annex V Garbage Disposal Limitations

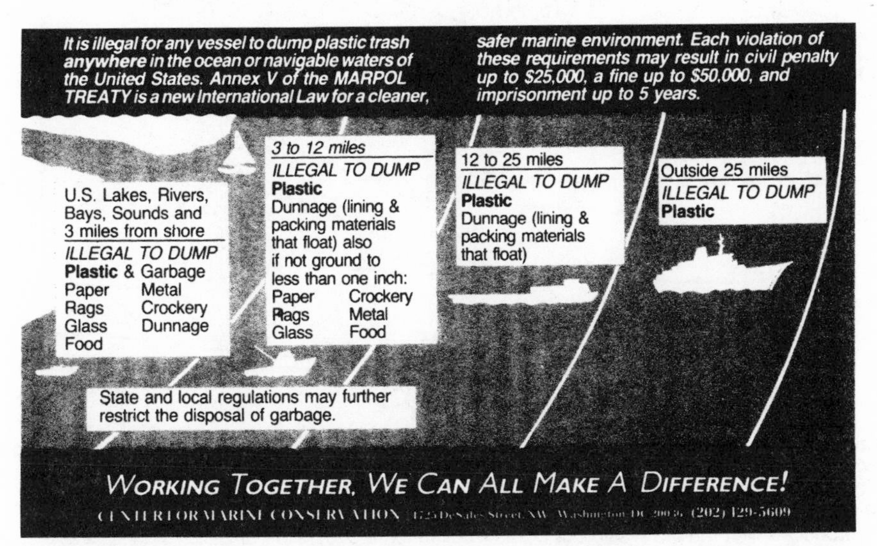

Source: *1992 International Coastal Cleanups Results.* Washington, DC: Center for Marine Conservation, p. 72.

Between 1973 and 1987, the amount of U.S. industrial waste dumped in the ocean annually declined steadily from about six million tons to well under one million tons. Permitted ocean dumping of industrial wastes ceased in September 1988.

Source: *United States of America National Report.* Prepared for the United Nations Conference on Environment and Development, Brazil, 1992. Washington, DC Council on Environmental Quality, 1992. p. 265.

Tables and Figures

Table 5 shows there was no decline in the production of toxic chemicals during the 1970s and 1980s (except for a slight decline in insecticides), only a slight decline in the number of pounds sold, and stability or an increase in the dollar sales amounts.

The use of herbicides increased steadily in the United States from 1964 until the early 1980s, when regulations regarding their use became stricter. The sharp decline in 1983 indicated in Figure 7 caused by these laws was followed, however, by continued increases as alternate substances were developed.

FIGURE 7 Use of Pesticides in the United States,1964 to 1984

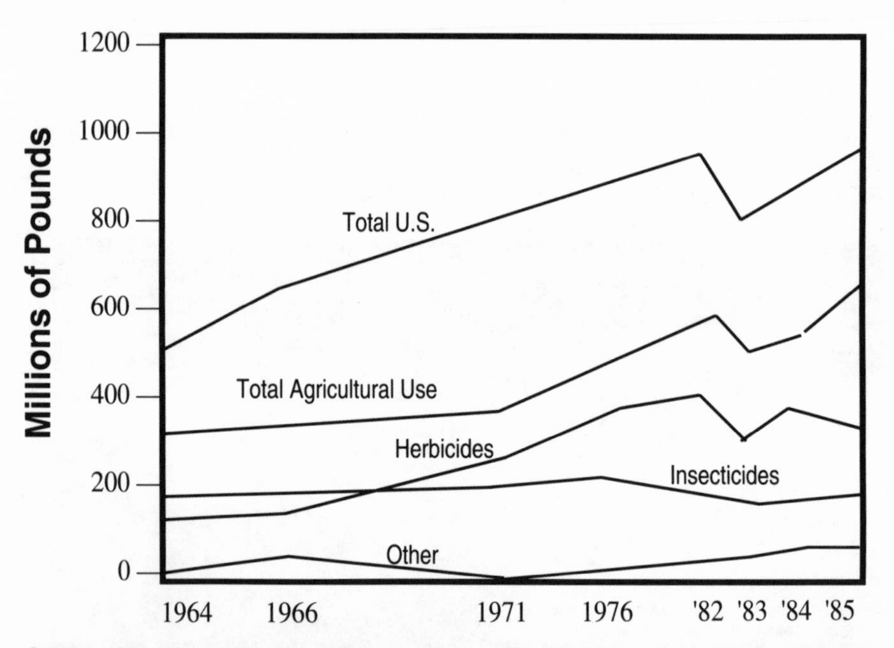

Source: Frits Van der Leeden, Fred L. Troise, and David Keith Todd, eds., *The Water Encyclopedia.* Chelsea, MI: Lewis Publishers, Inc., 1990. p. 585. (From: U.S. Environmental Protection Agency, Office of Pesticides and Toxic Substances, 1987, *Agricultural Chemicals in Ground Water: Proposed Pesticide Strategy.*)

TABLE 5 Synthetic Organic Pesticides—Production and Sales: 1970 to 1980

	Unit	1970	1975	1979	1980	1981	1982	1983	1984	1985	1986	1987	1988
Production Total	Mil. lb.	1,034	1,063	1,429	1,468	1,430	1,113	1,017	1,189	1,235	1,180	1,040	1,164
Herbicides	Mil. lb.	404	788	657	806	839	623	570	716	756	725	556	702
Insecticides	Mil. lb.	490	660	617	506	448	379	324	350	370	342	379	352
Fungicides	Mil. lb.	140	155	155	156	143	111	123	123	109	113	105	110
Production Value	Mil. $$	1,058	2,900	3,685	4,269	5,136	4,331	3,993	5,056	5,360	5,305	4,716	5,389
Sales, Total	Mil. lb.	881	1,317	1,369	1,406	1,291	1,147	1,017	1,108	1,022	904	911	935

Source: U.S. Department of Agriculture, Agricultural Stabilization and Conservation Service. *The Pesticide Review* (1978), and unpublished data. Based on data from U.S. International Trade Commission, *Synthetic Organic Chemicals,* annual.

TABLE 6 Selected Human Health and Environmental Effects from Toxic Chemicals

Chemical	Human Health Effects	Environmental Effects
Aldrin/dieldrin	tremors, convulsions, kidney damage	toxic to aquatic organisms, reproductive failure in birds and fish, bioaccumulation in aquatic organisms
Arsenic	vomiting, poisoning, liver and kidney damage	toxic to legume crops
Benzene	anemia, bone marrow damage	toxic to some fish and aquatic invertebrate
Cadmium	suspected causal factor in: tumors renal dysfunction, hypertension, arteriosclerosis, Itai-itai (weakened bones), emphysema	toxic to fish, bioaccumulates significantly i bivalve mollusks.
Carbon tetrachloride	kidney and liver damage, heart failure	
Chromium	kidney and gastrointestinal damage, respiratory complications	toxic to some aquatic organisms
Copper	gastrointestinal irritant, liver damage	toxic to juvenile fish & other aquatic organisms
DDT	tremors, convulsions, kidney damage	reproductive failure of birds and fish, bioac cumulates in aquatic organisms, biomagnifies in the food chain
Di-n-butyl phthalate	central nervous systemdamage	eggshell thinning in birds, toxic to some fi
Dioxin	acute skin rashes, systemic damage, mortality	bioaccumulates, lethal to aquatic organisr birds, and mammals
Lead	convulsions, anemia, kidney and brain damage	toxic to domestic plants and animals, biomagnifies to some degree in food chain
Methyl Mercury	irritability, depression, kidney and liver damage, Minamata disease	reproductive failure in fish species, inhibit growth and kills fish; biomagnifies
PCBs	vomiting, abdominal pain, temporary blindness, liver damage	liver damage in mammals, kidney damag and eggshell thinning in birds, suspected productive failure in fish
Phenols	effects on central nervous system, death at high doses	reproductive effects in aquatic organisms toxic to fish
Toxaphene	pathological changes in kidney & liver; changes in blood chemistry	decreased productivity of phytoplankton communities, birth defects in fish and birc toxic to fish and invertebrates

Source: Frits Van der Leeden, Fred L. Troise, and David Keith Todd, eds., *The Water Encyclopedia.* Chelsea, MI: Lewis Publishers, Inc., 1990. p. 622–23. (From: U.S. Environmental Protection Agency, National Water Quality Inventory, *1984 Report to Congress and The Conservation Foundation, State of the Environment 1982).*

Table 6 details the symptoms in humans and the impact on the environment of contamination by 15 common toxins. In many cases, the human health effects indicated are based on the results of animal tests.

Heavy Metals

Facts and Statistics

The annual world production of mercury reached a peak of 10,600 tons in 1971, falling to a little over 6,000 tons in 1987.

Source: Robert B. Clark, *Marine Pollution.* 2nd rev. ed. New York: Oxford University Press, 1989. p. 83.

The total global input of mercury to the sea from volcanic activity and weathering of mercury-bearing ores is at least 5,000 tons per year, but may be as high as 25,000 tons. A further 5,000 tons per year are added from the use of mercury and mercurial compounds in industry and agriculture. An additional 3,000 tons are derived from burning fossil fuels, principally coal.

Source: Robert B. Clark, *Marine Pollution.* 2nd rev. ed. New York: Oxford University Press, 1989. p. 84.

Tables and Figures

Because estuaries are among the most fertile and productive ocean waters, the levels of heavy metals in these waters is very important. They receive the contaminants carried by rivers as well as those dumped directly into coastal waters. Estuaries with some of the highest concentrations of heavy metals are listed in Table 7, along with the amounts of metals present in the estuarine sediment in 1984, stated in parts per million.

Oil

Facts and Statistics

Experts estimate that each year between 3 million and 6 million metric tons of oil are discharged into the oceans from land- and sea-based sources. . . .

Source: World Resources Institute (WRI) and International Institute for Environment and Development (IIED), *World Resources 1987,* pp. 128–129; "The State of the Marine Environment," UNEP News, April 1988, p. 11.

TABLE 7 Trace Metals in Sediments from Selected Estuaries in the United States, 1984

Estuary	Chromium	Copper	Lead	Zinc	Cadmium	Mercury
Salem Harbor, MA	2296.67	95.07	186.33	238	5.87	1.19
Raritan Bay, NJ	181.00	181.00	181.00	433.75	2.74	2.34
Pamlico Sound, NC	79.67	14.13	30.67	102.67	0.33	0.11
Mobile Bay, AL	93.00	17.40	29.67	161.00	0.11	0.12
San Diego Harbor, CA	178.00	218.67	50.97	327.67	0.99	1.04
San Francisco Bay, CA	1466.67	160.71	67.39	501.66	0.51	0.25
Nisqually Reach, WA	118.07	13.33	24.57	105.33	0.68	0.32
Lutak Inlet, AK	58.27	26.67	15.90	180.33	0.96	0.24

Source: U.S. Department of Commerce/NOAA. National Oceanographic Survey, Ocean Assessments Division. National Status and Trends Program for Marine Environmental Quality. *Progress Report on Preliminary Assessment of Findings of the Benthic Surveillance Project,* 1984. Rockville, MD: U.S. Geological Survey, National Water Summary 1986.

The main source of sea-based oil is the shipping industry, which discharges roughly 1.5 million metric tons of oil into the oceans each year. About a ton of oil is discharged for every thousand tons transported by sea . . . The bulk of seaborne oil pollution results from washing tankers out with seawater and releasing oily ballast water into the ocean.

Source: Organization for Economic Cooperation and Development (OECD), *The State of the Environment 1985*. Paris: OECD, 1985. p. 74.

In 1989, the *Exxon Valdez* disaster off Alaska spilled more than 10 million gallons of oil and killed at least 23,000 migratory birds, 730 sea otters, and 50 birds of prey.

Source: "Wildlife Toll Still Rising in Alaska," *The Washington Post,* June 1, 1989, p. A3.

An estimated 10,000 spills take place in the U.S. each year . . . Crude oil tankers can now carry as much as 350,000 tons of oil.

Source: *Oil Spills: Just a Cost of Doing Business,* The Wilderness Society, May 1991.

As recently as 1948, no cargo ship weighed more than 26,000 dead weight tons. By 1973, there were more than 400 oil tankers of 200,000 or more tons, 2 of them of 447,000 tons. . . . They are built with skin-thin steel hulls and without the safety standards common to other types of ships.

Source: Richard A. Frank, "The Law at Sea," in *Time Magazine* (May 18) 1975. pp. 14–15.

The largest tanker in the world is called the *Hellas Fos.* It's 453 yards (414 meters) long, or as long as four and a half football fields.

Source: *Make a Splash: Care about the Ocean,* Brookfield, CT: The Millbrook Press, 1992. p. 17.

Tables and Figures

The major oil routes of the world are shown in Figure 8. These areas are the most likely to be the scene of a major oil spill, and all are subject to the impact of constant lesser spills. Oil pollution of the world's seas correlates closely to these tanker routes.

A gradual and uneven decrease in the number of oil spill incidents is revealed in Table 8, but no parallel decrease occurs in the total number of gallons of oil spilled. The number of spills that take place in U.S. waters alone are in the thousands, and the gallons are in the millions.

FIGURE 8 Major Oil Movements at Sea

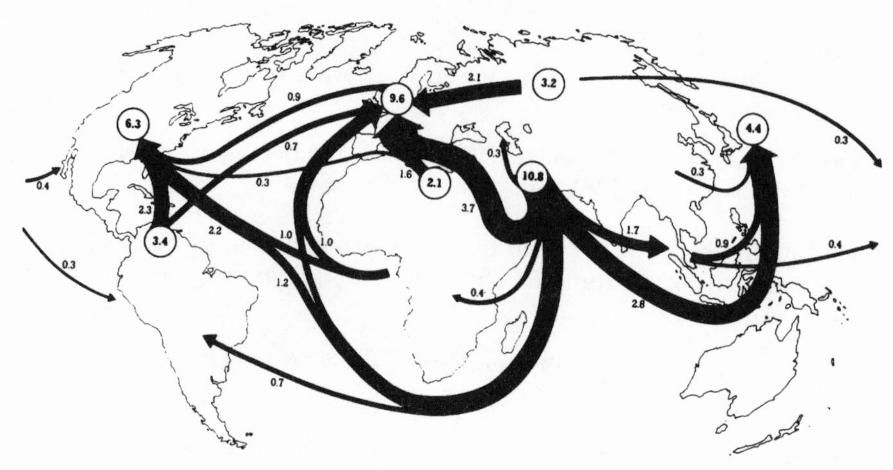

Source: Robert B. Clark, *Marine Pollution.* 2nd rev. ed. New York: Oxford University Press, 1989, p. 35.

Table 9 shows oil spills at a high in 1975, then declining in number and severity until the mid-1980s, when the figures began to rise again. Tonnage is revealing. Only 31 tanker casualties in 1989 resulted in the same tonnage lost as in 45 casualties in 1975. This reflects the extreme expansion in the size of the world's tankers.

Radioactive Materials

Facts and Statistics

The radioactivity of seawater from all natural radionuclides is about 750 disintegrations per minute/liter, 90 percent of which derives from potassium-40. Because of the long half-lives of many natural radionuclides, all organisms living in the ocean have been and will continue to be exposed to an essentially constant level of background radiation.

Source: Panel on Radioactivity in the Marine Environment, Committee on Oceanography of the National Research Council, *Radioactivity in the Marine Environment.* Washington, DC: National Academy of Sciences, 1971. p. 2.

There were 361 nuclear power plants in operation worldwide at the end of 1985.

Source: *Nuclear News,* February 1986.

TABLE 8 Oil Spills in and around U.S. Waters, 1970 to 1990

Year	Number	Volume
1970	3,710	15,250,000
1971	8,740	8,840,000
1972	9,930	18,810,000
1973	11,054	15,289,188
1974	12,083	15,739,792
1975	10,998	21,528,444
1976	11,066	18,517,384
1977	10,979	8,188,396
1978	12,174	11,035,890
1979	11,556	10,051,271
1980	9,886	12,636,848
1981	9,589	8,919,789
1982	9,416	10,404,646
1983	10,530	8,378,719
1984	10,089	16,254,974
1985	7,747	18,675,138
1986	6,543	4,668,441
1987	6,185	4,332,322
1988	6,788	6,640,851
1989	7,859	14,180,497

Source: *Statistical Abstract of the United States 1991.* p. 208. (From: Tanker Advisory Center, Inc., New York, NY, "Worldwide Tanker Casualty Returns," quarterly and from Council on Environmental Quality. *Environmental Quality, 22nd Annual Report.* Washington, DC: GPO, 1992. p. 341. based on U.S. Department of Transportation, United States Coast Guard, *Polluting Incidents in and around U.S. Waters,* COMDTINST M16450.2H. Washington, DC: DOT, 1989, with update).

TABLE 9 Worldwide Tanker Casualties: 1975 to 1990

Item	**Unit**	**1975**	**1980**	**1982**	**1983**	**1984**	**1985**	**1986**	**1987**	**1988**	**1989**	**1990**
Casualties	Number	906	(NA)	(NA)	(NA)	(NA)	340	451	408	456	528	151
Total Losses	Number	22	15	21	11	14	12	8	5	3	8	1
Deaths	Number	90	132	72	14	68	53	23	12	63	74	19
Oil Spills	Number	45	32	9	17	15	9	8	12	13	31	8
Amount	**1,000 tons**	**188.0**	**135.6**	**1.8**	**387.8**	**22.4**	**79.8**	**5.0**	**8.7**	**178.3**	**188.0**	**3.5**

Source: *Statistical Abstract of the United States 1991*. p. 638. (From: Tanker Advisory Center, Inc., New York, NY, "Worldwide Tanker Casualty Returns," quarterly.)

. . . there are now an estimated 585 nuclear-powered submarines in the world.

Source: K. A. Gourlay, *Poisoners of the Seas*. London and Atlantic Highlands, New Jersey: Zed Books Ltd., 1988. p. 180.

The known ocean disposal between the years 1946 and 1970 amounts to 94,673 curies of activity dropped from barges in depths of up to 12,000 feet in ten sites, including three off British Columbia . . . The preponderance of this material was simply sealed in 55-gallon or 80-gallon drums lined to 50 percent of their capacities with concrete as both shielding and ballast, and dropped at three principle sites, two of them off the Farallon Islands about 50 miles from San Francisco, and the third about 120 miles out in the Atlantic off the southern tip of New Jersey.

Source: Fred C. Shapiro, *Radwaste*. New York: Random House, 1981. p. 123.

Industry leaders know that by 1980 there were at least 2,535 tons (2301.27 metric tons) of spent fuel rods in temporary storage around the country. Unless something is done, there will be 108,000 tons (98,042.4 metric tons) by the year 2000.

Source: Ann E. *Weiss, The Nuclear Question*. New York: Harcourt Brace Jovanovich, 1981. p. 60.

. . . it is civilian nuclear power that has produced roughly 95 percent of the radioactivity emanating from waste in the world. In 1990, the world's 424 commercial nuclear reactors created some 9,500 tons of irradiated fuel, bringing the total accumulation of used fuel to 84,000 tons—twice as much as in 1985. The United States houses a quarter of this, with a radioactivity of more than 20 billion curies.

Source: Lester R. Brown and Christopher Flavin. *Vital Signs 1992: The Trends That Are Shaping Our Future*. Linda Starke, ed., Worldwatch Institute. NY, London: W.W. Norton & Co., 1992.

Tables and Figures

The amount of low-level radioactive waste accumulating at commercial sites across the nation grows in volume every year, as detailed in Table 10 (m3 indicates cubic meters). This is one of the reasons ocean dumping of these wastes is so appealing to the nuclear industry.

TABLE 10 Accumulated Volume and Radioactivity of Nuclear Waste in the United States, 1963 to 1990

Low-Level Wastes of Commercial Disposal Sites

Year	**Volume** (million m^3)	**Radioactivity** (million curies)
1963	0.008	0.042
1964	0.020	0.204
1965	0.034	0.273
1966	0.049	0.355
1967	0.071	0.428
1968	0.091	0.529
1969	0.112	0.687
1970	0.138	0.855
1971	0.169	2.000
1972	0.208	2.287
1973	0.255	2.732
1974	0.309	2.754
1975	0.367	3.040
1976	0.442	3.268
1977	0.514	3.765
1978	0.593	4.383
1979	0.676	4.539
1980	0.768	4.547
1981	0.852	4.483
1982	0.929	4.568
1983	1.007	4.732
1984	1.083	4.954
1985	1.160	5.282
1986	1.213	5.509
1987	1.265	4.924
1988	1.306	4.793
1989	1.352	5.284
1990	1.384	5.349

Source: *Environmental Quality, 22nd Annual Report.* Washington, DC: GPO, 1992. p. 337. (from: U.S. Department of Energy, Integrated Data Base for 1991: Spent Fuel and Radioactive Waste Inventories, Projections, and Characteristics. Washington, DC: DOE, 1991).

TABLE 11 Accumulated Volume and Radioactivity of High-Level Nuclear Waste in the United States, 1980 to 1990

DOE/-Defense Sites			Commercial		
Year	Volume (mill. m^3)	Radioactivity (bill. curies)	Year	Volume (thousand m^3)	Radioactivity (mill. curies)
1980	0.30	1.31	1980	2.20	34.20
1981	0.31	1.58	1981	2.20	33.40
1982	0.34	1.32	1982	2.20	32.70
1983	0.35	1.25	1983	2.20	31.90
1984	0.36	1.40	1984	2.20	31.20
1985	0.36	1.47	1985	2.20	30.40
1986	0.36	1.42	1986	2.20	29.70
1987	0.38	1.28	1987	2.20	29.20
1988	0.38	1.18	1988	2.10	28.70
1989	0.38	1.08	1989	2.40	27.90
1990	0.40	1.02	1990	1.20	27.30

Source: *Environmental Quality, 22nd Annual Report*. Washington, DC: GPO, 1992. p. 337. (from: U.S. Department of Energy, Integrated Data Base for 1991: Spent Fuel and Radioactive Waste Inventories, Projections, and Characteristics. Washington, DC: DOE, 1991).

High-level radioactive wastes, which include spent fuel rods, for instance, are accumulating at both commercial sites (like nuclear power facilities) and at defense plants (such as Rocky Flats in Denver, Colorado), where nuclear weapons have been manufactured. Note that Table 11 lists waste volume in *thousands* of cubic meters for commercial sites, but in *millions* for Department of Energy (DOE) and defense sites.

The peak seen in 1964 in Figure 9 is probably due at least in part to the effect of a U.S. atomic satellite entering the Earth's atmosphere 28 miles above the Indian Ocean, as well as to the fact that the test ban treaty had not yet gone into effect.

FIGURE 9 Monthly Deposition of Strontium-90 in the Northern Hemisphere, Derived from Weapons Testing

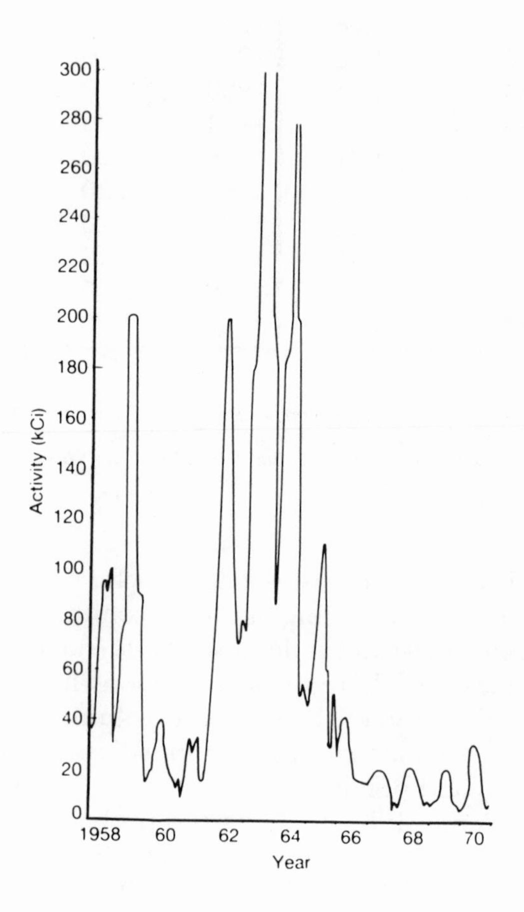

Source: Robert B. Clark, *Marine Pollution.* 2nd rev. ed. New York: Oxford University Press, 1989. p. 135.

Points of View

While statistics can reveal the extent of a problem, the words of persons directly involved provide insight and perspective that numbers often lack.

What Is Marine Pollution?

In a special section in the 1963 *World Book Yearbook* called "The Living Ocean," there is no mention of marine pollution except for a single sentence of a letter attributed to President John F. Kennedy:

> Many phases of human life are influenced by the sea. It is an abundant source of food and minerals. It shapes the climate in which we live. The sea also serves as a disposal area for a large variety of waste materials . . .
> **Source:** 1963 *World Book Yearbook* special section: "The Living Ocean."

At a UNESCO meeting held in Paris in 1967, the United States contributed the following "Conception of the Ocean as Resource":

> The basic thrust of sea pollution control programmes should be to reorganise the ocean as a productive resource providing a variety of goods and services. . . .
> **Source:** *The Intergovernmental Oceanographic Commission Report on Oceanic Pollution.* UNESCO-AVS/9/87 Paris, August 1967.

A Report written by the Secretary-General of the United Nations in 1968 on marine science and technology presents the following perception of marine pollution:

> The inshore marine environment must be protected against deterioration resulting from the discharge of municipal and industrial wastes. This environment has the capacity to receive a certain amount of waste discharge without damage to its other uses and in fact a valuable and legitimate use of the near shore marine environment is as a diluting and assimilating environment for waste materials, provided these are introduced within the capacity of the environment. By capacity is meant a rate of introduction which will not result in degradation from the standpoint of other uses, such as fishing and recreation . . .
> **Source:** *Marine Science and Technology: Survey and Proposals, Report of the Secretary-General.* In ECOSOC (UN), 1968. E/4487.

The first scientifically sanctioned definition of pollution was developed by the Joint Group of Experts on the Scientific Aspects of Marine Pollution (GESAMP) during the United Nations Conference on the Human Environment in 1972:

> The introduction by man, directly or indirectly, of substances or energy to the marine environment (including estuaries) resulting in deleterious effects such as: harm to living resources; hazards to human health; hindrance of marine activities including fishing; impairing the quality for use of seawater and reduction of amenities.
> **Source:** *GESAMP III/19, Report of the Third Session.* Reprinted in H. Windom: *GESAMP, Two Decades of Accomplishments,* International Maritime Organisation, 1991. p. 2.

Explorer Thor Heyerdahl helped introduce a popular concept of marine pollution to the general public following his second sea voyage, and his subsequent talks and articles on the changes he witnessed:

> Man is not the first polluter. Since the morning of time nature itself has been a giant workshop, experimenting, inventing, decomposing, and throwing away waste: the incalculable billions of tons of rotting forest products, decomposing flesh, mud, silt and excrement. If this waste had not been recycled, the ocean would long since have become a compact soup after millions of years of death and decay, volcanic eruptions, and global erosion. Man is not the first large-scale producer, so why should he become the first disastrous polluter?
>
> Man has imitated nature by manipulating atoms, taking them apart and grouping them together in different compositions. Nature turned fish into birds and beasts into man. It found a way to make fruits out of soil and sunshine. It invented radar for bats and whales, and shortwave transceivers for beetles and butterflies. Jet propulsion was installed on squids, and unsurpassed computers were made as brains for mankind. Marine bacteria and plankton transformed the dead generations into new life. The life cycle of spaceship earth is the closest one can ever get to the greatest of all inventions, *perpetuum mobile*—the perpetual-motion machine. And the secret is that nothing was composed by nature that could not be recomposed, recycled, and brought back into service again in another form as another useful wheel in the smoothly running global machinery.
>
> This is where man has sidetracked nature. We put atoms together into molecules of types nature had carefully avoided . . . But with ever-increasing speed and intensity we now produce and discharge into the sea hundreds of thousands of chemicals and other products. They do not evaporate nor do they recycle, but they grow in numbers and quantity and threaten all marine life . . . A dead ocean means a dead planet.
> **Source:** Thor Heyerdahl, "How To Kill a Planet," in *Saturday Review,* Vol. 3, November 29, 1975. pp. 12–18.

Marine scientist Dr. Robert B. Clark summarizes the cost-benefit approach to marine pollution adopted by much of the scientific community during the 1980s:

> Generally the issues are not so clear cut and the decisions are more difficult. A more sophisticated approach is needed and, for this, a great deal of information is required:
> - What is the level of contamination in the area we are interested in?
> - What form does it take?
> - Where does it come from?
> - What happens to it?
> - What does it do to the plants and animals there?
> - If plants and animals are affected, does it matter?
> - To whom does it matter? What other interest is affected?
> - How much does it matter to them?
> - If it does matter, what can we do about it?
> - What do we do with the polluting material if it is not put into the sea?
> - Would the alternative be better or worse than putting it in the sea?
> - How much would it cost?
>
> **Source:** Robert B. Clark, *Marine Pollution.* New York: Oxford Press, 1989, p: 9.

Famed undersea explorer, marine scientist, and cinematographer Jacques-Yves Cousteau graphically and poignantly revealed the extent of human pollution of the ocean to a broad audience around the world:

> The ocean is the sewer for the entire planet; rain, streams, and rivers drag into the sea toxic particles from the smoke of the factories or of incinerators, lead compounds from the exhaust of automobiles, pesticides from farm and domestic sprays. Radioactive waste is dumped into the Atlantic. Any irreparable damage done to the water system is an unatonable crime; any human enterprise that might bring about such permanent scars must be withheld until it is proved harmless. The permanent damages that have already been done must be publicized as examples of mistakes to be avoided in the future.
>
> Mass slaughter of whales, incessant scraping of the North Sea's bottom with heavy trawl nets, killing of porpoises and dolphins in tuna purse nets, ravages of coral reefs by spearfishermen, oil drilling in unsafe offshore areas—these are examples of how a distorted image of progress can lead to a shameless rape of the sea. With increased production as the only goal of national governments, the responsibility to the environment and to future generations is abandoned. Today the word progress is used as a synonym for growth, and growth gets out of hand.
>
> To whom can we turn to obtain an unbiased evaluation of the risks we can accept, not only for ourselves but also for our children, our

grandchildren, and the hundreds and thousands of generations to come? The answer is obvious: the average citizen can rely only on his own judgement; his civic duty is to give utterance to that judgement by all means and as loud as he can.
Source: Jacques-Yves Cousteau, as quoted in: Robert Cahn, *Footprints on the Planet: A Search for an Environmental Ethic.* NY: Universe Books, 1978. p. 215.

Sewage

Roger Revelle, who was then the Director of Scripps Institution of Oceanography, made the following revealing statements as part of his Welcoming Address for the First International Conference on Waste Disposal in the Marine Environment, held at the University of California in 1959:

> I am convinced that if due care is exercised, based on adequate scientific knowledge and with generous respect for other peoples' uses of the oceans, very large amounts of wastes can be safely disposed of at sea . . . Biologists and oceanographers, when making recommendations about waste disposal into the marine environment, often tend to pile safety factors upon safety factors to arrive at quite unrealistic result . . . My welcoming advice, therefore, to this conference is: Be scientists, don't be special pleaders.
> **Source:** *Proceedings of the First International Conference on Waste Disposal in the Marine Environment,* University of California, Berkeley, July 1959.

A. M. Raum, who was the General Manager Emeritus, Los Angeles County Sanitation Districts and Chairman of the California State Water Pollution Control Board, made the following statements during a talk given at the First International Conference on Waste Disposal in the Marine Environment in 1959:

> No one can dispute the necessity for protection of recreational waters from sewage contamination, but the very admission of the justice of such protection raises the controversial question of whether or not certain areas of the sea near seaboard cities should be set aside for primary sewage effluent disposal purposes, with the public excluded from such areas. Those who can envisage the tremendous cost involved in building and operating complete treatment plants for all, or many, seaboard cities find great merit in such a limited ocean disposal plan. There is, however, opposition to such a course of action from those interested in water sports and other forms of recreation, a quite natural reaction to what is considered the despoliation of a natural resource. Until a matter such as this is settled by agreement or by law it probably will remain within the purview of

water pollution control authority and be viewed subjectively, case by case.

Source: A. M. Raum, *Proceedings of the First International Conference on Waste Disposal in the Marine Environment,* 1959.

In 1965, the concept of marine sewage disposal was just coming under scrutiny, as revealed in these cautionary comments made during the Third Session of the Advisory Committee on Marine Resources Research, held in Rome in 1965:

> It is generally known that pollution may lead to deterioration of environmental conditions and, in extreme cases, to complete destruction of a natural fishery resource. The popular idea that rivers are to be used as natural sewers is dying slowly, but it is still commonly accepted that the sea is an ideal place to deposit any unwanted by-products of civilisation and industrialisation. The Committee stresses, however, that great care must be exercised in the disposal of waste material if damage is not to be caused . . .
>
> With respect to industrial wastes there has been a development of national laws controlling or completely prohibiting their disposal on land or into fresh waters, and in some cases, into estuaries and territorial waters. This unfortunately, is contributing to a trend towards the uncontrolled dumping of such wastes on the high seas. The tendency is, for convenience, to dump these materials as near to the point of origin as existing limited legislation permits. Especially important among industrial wastes are those having organic components, including, in addition to pesticides of all kinds, chlorinated hydrocarbons, cyanides and solutions in which the solvent is chloroform . . .
>
> It is desirable that such pollutants as might continue deliberately to be introduced under control, should be disposed of in such a way that their effects on the fisheries resources are minimised, either through being introduced in zones where there is little or no water movement—if such zones exist—or, conversely, in areas where water movement will ensure wide dispersion and dilution of pollutants, or convey them to other areas where the harm they might do is minimised or localised.
>
> **Source:** *Report of the Third Session of the Advisory Committee on Marine Resources Research,* ACMRR(FAO) Third Session, Rome 1965.

In one of the last published arguments of this nature, W. Bascom, a scientist with the Woods Hole Oceanographic Institution, argues:

> Contrary to some widely held views, the ocean is the plausible place for man to dispose of some of his wastes. If the process is thoughtfully controlled, it will do no damage to marine life. . . . It is plausible

that a great deal of effort will be made to comply with laws that will do little to make the ocean cleaner.
Source: W. Bascom, "The Disposal of Waste in the Ocean." *Scientific American* Vol. 213, No. 2, August 1974. pp. 16-25.

The scientific community has still not reached a consensus on ocean disposal, but became far more cautious by the end of the 1980s:

The view that no wastes of any sort can be safely discharged into the sea—a few years ago regarded as an unreasonable and extreme position—is now much more widely accepted.
Source: Robert B. Clark, *Marine Pollution.* New York: Oxford Press, 1989. p. v.

Marine Debris

The problem of marine debris became too prevalent to ignore during the 1980s. Celebrities like Ted Danson (star of the television program *Cheers* and many movies) loaned their names and notoriety to the issue. Plastics in the marine environment began to be recognized nationally and internationally as a serious threat to marine life and to the health of the oceans:

Just look at the headlines—recreation beaches closed due to a dangerous level of trash on shore—commercial fishermen competing with their own lost gear—seals and seabirds strangled by the unbreakable rings of six-pack holders—sea turtles eating plastic bags, mistaking them for jellyfish—these are the results of our misuse of plastic, especially in how we dispose of it. Plastic doesn't disappear when we throw it away. We can't afford to ignore plastics and other non-degradable items just because we're finished with them. The ocean can't tolerate it. We must learn to be responsible plastics users.
Source: Ted Danson, in the Foreword to: Kathryn O'Hara, Suzanne Iudicello, and Rose Bierce, *A Citizen's Guide to Plastics in the Ocean: More Than a Litter Problem.* 3rd rev. ed. Washington, DC: Center for Marine Conservation, 1988. p. vii.

The Center for Marine Conservation has taken on the marine debris problem, which is primarily a plastics issue, as its major battle on behalf of the oceans:

The widespread presence of plastics in the oceans is a global problem that will require international cooperation to solve. But the roots of the problem stem from individual carelessness in disposing of a material that is part of our everyday lives. No one can point the finger

at a particular country, region, industry, or group as the major contributor to the problem. The responsibility is shared by us all.
Source: Kathryn O'Hara, Suzanne Iudicello, and Rose Bierce, *A Citizen's Guide to Plastics in the Ocean: More Than a Litter Problem.* 3rd rev. ed. Washington, DC: Center for Marine Conservation, 1988. p. 1.

There are three main approaches to solving the problem of marine debris: prevent it from entering the oceans, make it degradable so it will disappear if it does reach the seas, or clean it up when it washes ashore:

> ... But we must be careful not to view degradable plastics as a panacea to the debris problem. It is neither desirable nor feasible to make all plastics degradable. In the first place, degradability does nothing to discourage the practice of using the ocean as a garbage can. If consumers and seafarers thought all plastics were degradable, they would have no incentive to carry them away from beaches or to stop dumping wastes overboard. Secondly, not all plastic products lend themselves to degradable applications. Degradability should be considered primarily for those plastic products that pose the greatest threat to marine resources.
>
> Ultimately, the solutions to the problems of plastic debris depend on continued cooperation among industry, policy makers, and the general public. At no time in history has attention to the marine pollution problem been greater. We must use this momentum to confront these problem areas and eliminate their contributions to the marine debris problem.
>
> **Source:** Kathryn O'Hara, Suzanne Iudicello, and Rose Bierce, *A Citizen's Guide to Plastics in the Ocean: More Than a Litter Problem.* 3rd rev. ed. Washington, DC: Center for Marine Conservation, 1988. p. 85.

Toxic Chemicals

Both the environmental dangers and the limitations of DDT as an insecticide were recognized during the 1940s, but warnings like the following by scientists with the U.S. Fish and Wildlife Service were ignored:

> From the beginning of its wartime use as an insecticide the potency of DDT has been the cause of both enthusiasm and grave concern. Some have come to consider it a cure-all for insect pests; others are alarmed because of its potential harm. The experienced control worker realizes that DDT, like every other effective insecticide or rodenticide, is really a two-edged sword; the more potent the poison, the more damage it is capable of doing. Most organic and mineral poisons are specific to a degree; they do not strike the innumerable

animal and plant species with equal effectiveness; if these poisons did, the advantage of control of undesirable species would be more than offset by the detriment to desirable and beneficial forms. DDT is no exception to this rule. Certainly such an effective poison will destroy beneficial insects, plants, and wildlife.

Although many more investigations are needed in all these fields, it seems that the most pressing requirement is a study to determine the effects of DDT as applied to agricultural crops on the wildlife and game dependent upon an agricultural environment. About 80 percent of our game birds, as well as a very high percentage of our non-game and insectivorous birds, and mammals are largely dependent on an agricultural environment. In such places application of DDT will probably be heavy and widespread; therefore it is not improbable that the greatest damage to wildlife will occur there.

Because of the sensitivity of fishes and crabs to DDT, avoid as far as possible direct applications to streams, lakes and coastal bays. Whenever DDT is used, make careful before and after observations of mammals, birds and fishes, and other wildlife.

Source: Clarence Cottam and Elmer Higgins, of the Fish and Wildlife Service, "DDT and its Effect on Fish and Wildlife," in the *Journal of Economic Entomology*. Vol. 39, February 1946, pp. 44–52.

Edward D. Goldberg, Scripps Institution of Oceanography marine chemist known for his study of DDT in the marine environment, published the following strong statements in 1971:

The more the problems are studied, the more unexpected effects are identified. In view of the findings of the past decade, our prediction of the potential hazard of chlorinated hydrocarbons in the marine environment may be vastly underestimated. The Panel makes the following recommendations, which will be developed and expanded in the remainder of the report:

- A massive national effort should be made immediately to effect a drastic reduction of the escape of persistent toxicants into the environment, with the ultimate aim of achieving virtual cessation in the shortest possible time.
- Programs should be designed both to determine the rates of entry of each pollutant into the marine environment and to make baseline determinations of the distribution of the pollutants among the components of that environment. These should be followed by a program of monitoring long-term trends in order to record progress and to document possible disaster.
- The laws relating to the registration of chemical substances and the release of production figures by governments should be examined and perhaps revised in light of evidence of environmental deterioration caused by some of these substances.

Source: Edward D. Goldberg, et al., *Chlorinated Hydrocarbons in the Marine Environment.* Washington, DC: National Academy of Sciences, 1971. p. 2.

This editorial published in England in 1971 characterizes the times:

> As the public outcry against pollution goes up in most western countries, Governments are being nudged into a show of action and industrial firms appear to exercise greater discretion in the disposal of their waste chemicals and other noxious refuse. Not that any noticeable impression is being made on a menace that increases inexorably with the growth of industrialization. The effect of such restrictions within national territory, however, is to encourage resort to the open seas as a free garbage pit for the disposal of industrial detritus. Only in the most glaring cases, such as the highly publicized *Stella Maris* with its 600 tons of chemical waste, is any positive action taken, and then only as the result of diplomatic intervention by the countries most concerned. Judging by the difficulty of securing compliance even with navigation rules in the world's busiest shipping lanes in the English Channel, the chances of achieving international agreement to limit the pollution of the wider seas are remote. The best hope lies in regional co-operation, in the appeal to national self-interest of the coastal countries most affected.
> **Source:** London *Daily Telegraph,* July 24, 1971, headlined "Marine Dustbin."

The U.S. House Committee on Public Works and Transportation held a Coastal Pollution Hearing on May 9 and 10, 1990. This is what the Committee learned:

> [U.S.] coastal waters are . . . polluted by poison runoff from city streets, farms, building sites, parking lots, and other common uses of land. In fact, this type of pollution causes at least half of the nation's water pollution. According to figures released by the EPA in 1989, of over 17,000 heavily polluted waters nationwide, less than 600 can be tied exclusively to point sources such as factories and sewage plants. The remaining waters are polluted, at least in part, by poison runoff. It is the leading source of many dangerous contaminants.
> **Source:** U.S. House Committee on Public Works and Transportation. *Coastal Pollution.* Hearing, 9–10 May, 1990. Washington, DC: GPO, 1991. p. 330.

Heavy Metals

The world reaction to the mercury poisoning in Minamata, Japan varied. This example from Sweden, as related by Dr. Sebastian A. Gerlach, a German marine scientist, reveals the complexity of the issue:

> When it became known how dangerous mercury is for man and to what extent high mercury concentrations in fish may accumulate

from polluted waters, governments had to react with legislation. When it became known from Japan that Minamata Bay fish had concentrations of 50 mg/kg, the Swedish health Ministry determined 0.5 mg/kg to be the limit of mercury concentration tolerable in fish for human consumption. Such regulation would provide a safety margin of 100 which is common practice in environmental regulations. But then two facts became known: first it was discovered that fish in Lake Vaumlautnern in Sweden exceeded this limit, and therefore the limit was elevated to 1 mg/kg. Then, at a public hearing, it became known that information from Japan related to dry weight, whilst Swedish regulations related to wet weight. This means that polluted fish (about 80% water content) from Minamata Bay had a mercury concentration of only 10 mg/kg, on weight basis. The safety margin was down from 1:100 to bare 1:10.
Source: Sebastian A. Gerlach, *Marine Pollution: Diagnosis and Therapy*. Berlin/New York: Springer-Verlag, 1981. p. 147.

Oil

The blow-out of an offshore oil well in the Santa Barbara channel, just off the coast of California, spewed at least 200,000 gallons into the sea and inspired insights such as the following:

If you looked carefully at what was happening in Santa Barbara, looked beyond the oily beaches and the dying birds and the anti-oil slogans, you could see a number of important issues raising their heads. There were legal, social, political, economic and philosophical questions to be answered, not only for Santa Barbara but for all America. The problem was so enormous that it was necessary to confront it one piece at a time; but then the partial solutions were often incompatible, and you were left once again with the discouragingly intractable whole. The perhaps irreconcilable differences between local and national interests, our dream of the "good life" for all and the environmental costs of obtaining it, our passion for exploitation and growth and our longing for peace and beauty.
Source: Carol E. and John S. Steinhart, *Blow Out: A Case Study of the Santa Barbara Oil Spill*. Belmont, CA: Duxbury Press, 1971. p. 106

Conferences on oil pollution prevention and cleanup proliferated during the 1970s. The following statement was made during the Joint Conference on Prevention and Control of Oil Spills in 1973:

It is axiomatic that the best time to begin the pollution prevention program is during the design phase of a new system or facility. This is the point at which we are able to exercise maximum control over

the selection of all the equipment and materials which will go into the facility. Modern technology and innovative thinking can be applied; necessary pollution abatement facilities and equipment and appropriate monitor, alarm and shutdown systems can be incorporated into the initial design. Properly designed and installed, a new system or facility might be expected to function for years with low maintenance and repair costs and minimum adverse effects on the environment. . . . [However], generally we find ourselves operating systems that have been in service for many years. The systems were designed to fit a condition that no longer exists, systems that have been subjected to corrosion, abrasion, and the general wear and tear of many years of service. This situation is not always to our liking, but it is where we usually are and where we usually have to start our program.
Source: Frank C. Folger, Jr., in a speech given at the Joint Conference on Prevention and Control of Oil Spills, March 1973.

The marine pollution focus moves from oil to substances even more hazardous, and to the long-term impact of ocean dumping:

There has also been a change of "fashions" in marine pollution. What were seen as major problems a few years ago have now been superseded. Oil pollution, for some 20 years seen as the greatest threat in the marine environment, now has little prominence. Eutrophication, once a problem in some lakes but regarded as of minor importance in the sea, has recently become a major preoccupation in the North Sea and several other sea areas.
Source: Robert B. Clark. *Marine Pollution.* New York: Oxford Press, 1989. p. iii.

Referring to the financial impact of the *Exxon Valdez* oil spill, Lawrence Rawi, Chairman of the Exxon Corporation said the $1 billion settlement and the $1 million criminal fine:

". . . would not have a significant effect on our earnings."
Source: *Oil Spills: Just a Cost of Doing Business,* Washington, DC: The Wilderness Society, May 1991.

The economics of oil spills is revealed during the aftermath of the *Exxon Valdez* spill in Alaska:

An estimated 10,000 spills take place in the U.S. each year, yet the Department of Justice collected less than $30 million in 1990 for all environmental violations. In many cases, as with the *Exxon Valdez* settlement, civil suits are tax-deductible.
Source: *Oil Spills: Just a Cost of Doing Business.* Washington, DC: The Wilderness Society, May, 1991.

Radioactive Materials

How much is too much? This question, addressed as early as 1971 by the Panel on Radioactivity in the Marine Environment (a National Research Council Committee), still has not been satisfactorily answered.

> . . .Man may be exposed to ionizing radiation from seawater in various ways. Swimming, walking on beaches, and handling contaminated fishing gear are some of the ways, but none is as important as the ingestion of seafoods. The question, "How much seafood can be eaten safely?" then arises and leads to the concept of an "acceptable dose." The acceptable-dose concept implies that exposure to ionizing radiation from any source entails some risk of a biological effect; therefore, a dose is considered acceptable only if the benefits are greater than the risks and the risks are acceptable both to the individual and to society as a whole.
> **Source:** Panel on Radioactivity in the Marine Environment, Committee on Oceanography of the National Research Council. *Radioactivity in the Marine Environment.* Washington, DC: National Academy of Sciences, 1971. p. 4.

The difficult subject of how to determine how much radiation is too much enters marine dumping discussions:

> In calculating recommended limits, it will be evident that the ICRP [International Commission on Radiological Protection] cannot use scientific judgements alone, and value judgements are necessary. ICRP has aimed to set a 'permissible' or 'acceptable' upper limit such that the risks to individuals from the radiation hazard are similar or lower than the risks commonly accepted in everyday life. This approach assumes, however, that the practice which gives rise to the risk is commonly 'accepted', and that the health effects are also comparable. . . . no practice shall be adopted unless its introduction produces a net positive benefit . . . all exposures shall be kept as low as reasonably achievable (the ALARA principle), economic and social factors being taken into account.
> **Source:** F. G. T. Holliday (Chairman), *Review of the Disposal of Radioactive Waste in the North East Atlantic.* London: HMSO, 1984. p. 50.

There is some dissension regarding the makeup of the International Commission on Radiological Protection:

> Membership in the ICRP [International Commission on Radiological Protection] is highly selective and controlled. Prospective membership must be recommended either by current ICRP members or by members of the International Congress of Radiology and then

approved by the ICRP International Executive Committee. *Through this structure, participation in standard setting has been dominated by colleagues from the military, the civilian nuclear establishment and the medical radiological societies, who nominate one another ... People in all these categories have a vested interest in the use of radiation and depreciation of the risks in its use.* There is the added problem of military secrecy in many countries, including the USA, about radiation health effects, since these are the results of a nuclear bomb. This again limits the pool of 'experts' available to ICRP. There is no independent body, not even the World Health Organisation, which can place a person on the ICRP. *It is, in every sense of the term, a closed club and not a body of independent scientific experts.*
Source: Rosalie Bertell, 1985, as quoted in: K. A. Gourlay, *Poisoners of the Seas.* London and Atlantic Highlands, New Jersey: Zed Books Ltd., 1988. p. 178.

The problem of how to safely dispose of radwaste was recognized before nuclear power plants proliferated:

... there should be no commitment to a large programme of nuclear fission power until it has been demonstrated beyond reasonable doubt that a method exists to ensure safe containment of long-lived highly radioactive waste for the indefinite future.
Source: RCEP (Royal Commission on Environmental Pollution), *6th Report: Nuclear Power and the Environment.* UK, 1976.

Our Changing Perceptions of Marine Pollution

This early conference on marine pollution put into words the core issue that still divides the business community and environmentalists:

What is at issue is the nature of the compromise that must be effected between, on the one hand, advances in science and technology leading to the raising of living standards not only of the developed countries but also of the developing countries and, on the other hand, the preservation of man's environment.
Source: *Proceedings of the Rome Conference on Oil Pollution of the Sea,* October 1968. Paper No. 17. Page 2.

Perhaps because Europe is so much more densely populated, marine pollution was recognized as a serious problem there before it was given much attention in the United States:

Man has hardly begun to put the sea to useful purposes, but he has already started to rob himself of the last and greatest source of raw materials by poisoning it!
Source: Cod-Christian Troebst, *Conquest of the Sea.* Trans. from German by Brian C. and Elsbeth Price. New York: Harper & Row, 1962. p. 209.

Walter Hickel, Secretary of the Interior, offered this testimony in 1969 during early consideration of the Coastal Zone Management Act:

> The National Estuarine Pollution Study concludes that our estuarine areas are seriously polluted, and that unwise use of the lands and waters of our estuarine zones not only contributes to this pollution but is rapidly destroying valuable natural resources. While the statutory directive was to study the estuarine zones, the findings concluded that the management problems of our estuaries relate directly to the entire coastal zone and that any management system must deal with the Coastal Zone and its entirety.
> **Source:** Walter Hickel, Secretary of the Interior, on December 3, 1969.

It is notable that the words *pollution, prevention,* and *preservation* do not form part of this important statement of intent issued by the Commission on Marine Science in 1969:

> A conviction that the time had arrived for this country to give serious and systematic attention to our marine environment and to the potential resources of the oceans. A national determination to take the steps necessary to stimulate marine exploration, science, technology, and financial investment on a vastly augmented scale.
> **Source:** Commission on Marine Science, Engineering and Resources (COMSER), *Our Nation and the Sea.* 1969.

The discussion of the ethics of marine pollution that began during the early 1990s is foreshadowed by this insightful statement contained in a report published in Sweden in 1967:

> Technology has advanced without due attention to pollution, its economic aspects in particular. Calculations of profit and loss—including all costs—have never been made. Very often only the immediate profits of a technical achievement are considered while the costs of accompanying damage are neglected.
> **Source:** Report of the Swedish Royal Commission on Natural Resources, Nov. 1967, as documented in M. M. Sibthorp in *Oceanic Pollution: A Survey and Some Suggestions for Control.* London: Thorney House, Smith Square, 1969. p. 1.

One of the first and most clearly articulated descriptions of the interconnectedness of the biosphere was written by M. M. Sibthorp in a paper for the Committee on Sea-Bed Resources in 1969. He cites the *Report of the Swedish Royal Commission on Natural Resources,* November 1967 (Stockholm) as the source of several of the concepts.

> . . . it is quite impossible to separate marine pollution from pollution of the biosphere in general. Pollution of the biosphere is one problem with many facets. Because of the transfer of pollutants between atmosphere, water and soil, the problem of pollution must always be regarded as a whole, and, moreover, it is becoming a very urgent one since such pollution is, without doubt, increasing. . . . [F]or the first time in history man has the power to destroy his environment and to do so comparatively rapidly. If he fails to control the technology he has evolved there is increasing evidence to show that this result will inevitably follow.
> **Source:** M. M. Sibthorp, *Oceanic Pollution: A Survey and Some Suggestions for Control.* Stockholm: Committee on Sea-Bed Resources, Parliamentary Committee for World Government, 1969.

This imaginary scenario, written and published by biologist and ecologist Paul Ehrlich in 1969, helped raise the American consciousness regarding the delicate eco-balance of the planet:

> It was clear by 1975 that the entire ecology of plankton, the tiny animals which eat the phytoplankton [microscopic plants] community led inevitably to changes in the community of zooplankton. These changes were passed on up the chains of life in the oceans to the herring, plaice, cod and tuna. As the diversity of life in the ocean diminished, its stability also decreased . . . The end of the ocean came late in the summer of 1979, and it came even more rapidly than the biologists had expected. There had been signs for more than a decade.
> **Source:** Paul Ehrlich, "Ecocatastrophe!" *Ramparts Magazine,* September 1969. pp. 24-28.

In response to questions asked by reporters regarding the exploitation of argonite on the sea bed of the Bahamas, a Dillingham Corporation (Honolulu) executive said:

> We thought we had something marketable and we went ahead and sold it. Maybe some fish were disturbed, but they could probably find some place elsewhere.
> **Source:** Burnell and von Simeon, eds., *Pacem in Maribus: Ocean Enterprises.* Santa Barbara, CA: The Center for the Study of Democratic Institutions, 1970. p. 6.

Undersea explorer, marine scientist, and cinematographer Jacques-Yves Cousteau here attempts to impress upon the U.S. Senate Subcommittee on Oceans and the Atmosphere the importance of saving the earth's oceans:

> The sea is threatened. We are facing the destruction of the ocean by pollution and by other causes. My role in this gigantic enterprise is

> only that of a witness, a modest witness, who has only one valuable thing to testify about and it is, I think, a unique quality of experience—underwater searching with companions for more than thirty years.
>
> We believe that the damage done to the ocean in the last twenty years is somewhere between 30 percent and 50 percent, which is a frightening figure. And this damage carries on at very high speed—to the Indian Ocean, to the Red Sea, to the Mediterranean, to the Atlantic. Our latest observations in the Pacific Ocean, in Micronesia and New Caledonia and in the Fiji Islands, are even more frightening.
>
> **Source:** Commandant Jacques-Yves Cousteau, testimony before the U.S. Senate Subcommittee on Oceans and the Atmosphere in 1971.

This statement echoes a similar quote attributed to Albert Einstein ("Man has lost the capacity to foresee and forestall. He will end by destroying the earth."):

> Man's technological ingenuity and scientific ability exceed his political foresight and the capacity to adapt his moral, political, and legal conduct to the technological changes that have revolutionized contemporary society.
>
> **Source:** R. Shinn, *The International Politics of Marine Pollution Control,* 1974.

Explorer Thor Heyerdahl helped introduce the general public to the concept of the ocean as an essential organ of a living planet:

> Since the ancient Greeks maintained that the earth was round and great navigators like Columbus and Magellan demonstrated that this assertion was true, no geographical discovery has been more important than what we all are beginning to discover today: that our planet has exceedingly restricted dimensions. There is a limit to all resources. Even the height of the atmosphere and the depth of soil and water represent layers so thin that they would disappear entirely if reduced to scale on the surface of a common-sized globe.
>
> The correct concept of our very remarkable planet, rotating as a small and fertile oasis, two-thirds covered by life-giving water, and teeming with life in a solar system otherwise unfit for man, becomes clearer for us as the progress of moon travel and modern astronomy. Our concern about the limits to human expansion increases as science produces ever more exact data on the measurable resources that mankind has in stock for all the years to come.
>
> **Source:** Thor Heyerdahl, "How To Kill a Planet," *Saturday Review,* Vol. 3, November 29, 1975. pp. 12–18.

This remarkable statement was made by Barbara Ward, of the International Institute for Environment and Development, during the 1970s:

> . . . Regeneration is a special property of water and it is exercised on an inconceivably massive and majestic scale by the oceans, the healers and cleansers of the whole terrestrial globe. But the system is nonetheless an organic system. As such it is still capable of death. . . .
>
> . . . One of the profound problems posed by Nature's "thresholds" is that the approach to the point of no return may give few if any danger signals. Red lights do not flash on in the deeps as one more species—of whose role in the total ecosystem we are completely ignorant—heads for death. Perhaps its survival did not matter. But perhaps it was the catalyst of an entire feeding system on which other species depend. So another threshold approaches. But again, there is no blowing of sirens or rockets in the sky. One more piece of the biosphere falls away. We do not even know what we are losing. Even if we did, it would be too late.
>
> **Source:** Barbara Ward, *Towards an Environmentally Sound Law of the Sea.* Washington, DC: International Institute for Environment and Development, 1976. p. ii.

A unique perspective on marine polluters:

> Applying the principle of universality to pollution, this study argues that since the resources of the sea are for use and enjoyment by the wider international community. . . . and since pollution may in fact destroy it, or indiscriminately endanger the health of consumers, polluters ought to be declared a common enemy of that community—a modern *hostes humani generis* [enemy of all, as pirates were deemed].
>
> **Source:** C. Odidi Okidi, *Regional Control of Ocean Pollution: Legal and Institutional Problems and Prospects.* The Netherlands: Sijthoff & Noordhoff, 1978. p. 268.

This statement marks the beginning of the ethics discussion that is currently taking place among environmentalists in the United States and abroad:

> If we choose to adhere strictly to the objective of maximizing the short-term rewards of the present generation, there are in fact no long term trade-offs to be made. . . . The question about use or non-use of DDT, for example, would be easily resolved. The fact that 1.3 billion people today can live in safety from malaria due to DDT would strongly outweigh the costs—for instance, in the form of inedible fish—inflicted upon future generations through our continued use of the chemical . . . We feel the moral and ethical leaders of our societies should adopt the goal of increasing the time-horizon implicit in man's activities—that is, introducing the longer term

function which maximizes the benefit of those living today, subject to the constraint that it does not decrease the economic and social options of those who will inherit this globe, our children and grandchildren . . .
Source: Donella Meadows and Jorgen Randers (MIT Limits to Growth Team), "The Carrying Capacity of Our Global Environment: A Look at Ethical Alternatives," as cited in Cahn, Robert, *Footprints on the Planet: A Search for an Environmental Ethic.* NY: Universe Books, 1978. p. 217.

This simple statement sums up the single most important shift in our perception of the environment in the past several decades. Our understanding of the planet as having a single, integrated biosphere is paving the way for increased concern about marine pollution issues.

Finally we are learning that the entire biosphere is interconnected, that we can't change one thing without changing another. This, in our view, is the Central Great Idea of ecology.
Source: Robert M. Hazen and James Trefil, *Science Matters: Achieving Scientific Literacy,* New York: Doubleday, 1992, p.262.

In 1991, the Joint Group of Experts on the Scientific Aspects of Marine Pollution (GESAMP) published *Review of the State of the Marine Environment,* a booklet that updates a similar 1972 publication. In it, this international group of scientists declares that:

. . . at the end of the 1980s, the major causes of immediate concern in the marine environment on a global basis are coastal development and the attendant destruction of habitats, eutrophication, microbial contamination of seafood and beaches, fouling of the seas by plastic litter, progressive build-up of chlorinated hydrocarbons, especially in the tropics and the subtropics, and accumulation of tar on beaches. However, concerns may differ from region to region, reflecting local situations and priorities. Furthermore, throughout the world, public perception may still accord greater importance to other contaminants such as radionuclides, trace elements and oil. These were highlighted in the 1982 GESAMP Review and are considered again in the present report, but we now regard them as being of lesser concern.

While no areas of the ocean and none of its principal resources appear to be irrevocably damaged, and most are still unpolluted, while there are encouraging signs that in some areas marine contamination is decreasing, we are concerned that too little is being done to correct or anticipate situations that call for action, that not enough consideration is being given to the consequences for the oceans of coastal development, and that activities on land continue with little regard for their effects in coastal waters. We fear, especially in view

of the continuing growth of human populations, that the marine environment could deteriorate significantly in the next decade unless strong, coordinated national and international action is taken now. At the national level in particular, the concerted application of measures to reduce wastes and to conserve raw materials will be essential. The efforts will be great and the costs high, but nothing less will ensure the continued health of the sea and the maintenance of its resources.
Source: GESAMP, "Review of the State of the Marine Environent," *GESAMP Report and Studies No. 39,* 1991. pp. 11–12.

In 1992, scientists affiliated with the Union of Concerned Scientists issued a dire warning to all nations:

Human beings and the natural world are on a collision course. Human activities inflict harsh and often irreversible damage on the environment and on critical resources. If not checked, many of our current practices put at serious risk the future that we wish for human society and the plant and animal kingdoms, and may so alter the living world that it will be unable to sustain life in the manner that we know. Fundamental changes are urgent if we are to avoid the collision our present course will bring about . . . Destructive pressure on the oceans is severe, particularly in the coastal regions which produce most of the world's food fish. The total marine catch is now at or above the estimated maximum sustainable yield. Some fisheries have already shown signs of collapse. Rivers carrying heavy burdens of eroded soil into the seas also carry industrial, municipal, agricultural, and livestock waste—some of it toxic . . . No more than one or a few decades remain before the chance to avert the threats we now confront will be lost and the prospects for humanity immeasurably diminished . . . We the undersigned, senior members of the world's scientific community, hereby warn all humanity of what lies ahead. A great change in our stewardship of the earth and the life on it, is required, if vast human misery is to be avoided and our global home on this planet is not to be irretrievably mutilated . . . Acting on this recognition is not altruism, but enlightened self-interest: whether industrialized or not, we all have but one lifeboat. No nation can escape from conflicts over increasingly scarce resources. In addition, environmental and economic instabilities will cause mass migrations with incalculable consequences for developed and undeveloped nations alike . . . A new ethic is required—a new attitude toward discharging our responsibility for caring for ourselves and for the earth. We must recognize the earth's limited capacity to provide for us. We must recognize its fragility. We must no longer allow it to be ravaged. This ethic must motivate a great movement, convincing reluctant leaders and reluctant governments and reluctant peoples themselves to effect the needed changes . . .
Source: Excerpted from *World Scientists' Warning to Humanity,* issued by the Union of Concerned Scientists in 1992.

5

Directory of Organizations

General Organizations

Advisory Committee on the Protection of the Sea (ACOPS)
57 Duke Street, Grovesner Square
London WIM 5DH
United Kingdom
Phone: 71-499-0704
71-493-3092 (FAX)
Secretary General

Founded in 1952 (as the Advisory Committee on Oil Pollution of the Sea), ACOPS then became the Advisory Committee on Pollution of the Sea until, in 1990, it changed its name again. Its purpose is to promote the preservation and protection of the world's seas from pollution by human activities. It currently focuses its efforts on land-based sources of marine pollution. ACOPS compiles statistics on pollution by oil and by dangerous substances.

PUBLICATIONS: *ACOPS Newsletter* (semi-annual), *ACOPS Survey* (annual), *ACOPS Yearbook* (annual).

Alaska Conservation Foundation (ACF)
430 West 7th Street, Suite 215
Anchorage, AK 99501
(907) 276-1917
(907) 274-4145 (FAX)
Jan Konigsberg, Executive Director

Since 1980, ACF has served as a central clearinghouse for environmental fund raising and projects to protect Alaska's fragile ecosystems. The projects it has underwritten include many related to the *Exxon Valdez* oil spill in Prince William Sound, on which it published weekly updates and maps on the cleanup.

PUBLICATIONS: *Grantseeker's Guide, Issues Updates* (periodic), annual report.

American Cetacean Society (ACS)
P.O. Box 2639
San Pedro, CA 90731
(213) 548-6279
(213) 548-6950 (FAX)
Patricia Warhol, Executive Director

ACS, the oldest whale conservation group in the world, keeps the public informed regarding hazards to the world's whale population. Although whaling is still a very real threat, equally dangerous today are the effects of pollution—especially marine debris and oil—on whales, dolphins, and porpoises. ACS lobbies in Washington to eliminate these hazards, creates education packets for classroom use, runs advertisements in national magazines, organizes whale-watching expeditions, and sends out two quarterly publications to members.

PUBLICATIONS: *Whalewatcher* (quarterly journal), *WhaleNews* (quarterly newsletter).

American Oceans Campaign
725 Arizona Avenue, Suite 102
Santa Monica, CA 90401
(310) 576-6162
(310) 576-6170 (FAX)
Ted Danson, President

Founded in 1987 by television and movie star Ted Danson, AOC works to preserve and protect marine ecosystems by educating the public and decisionmakers on the need to stop abusing the oceans. It was instrumental in banning driftnets in American waters, participated in the coalition against oil development in the Arctic National Wildlife Refuge, helped block lease sales for offshore oil and gas development, organized the National Coastal Caucus, worked to end the dumping of inadequately treated sewage and sludge off U.S. coasts, drafted beach protocols, and co-produced a number of conservation videos. AOC endeavors to raise the level of treatment for all sewage discharged into coastal waters and rivers, ban outer continental shelf oil and gas development, expand the number of national marine sanctuaries, protect fisheries, and create a network of international citizen's groups dedicated to preserving the world's oceans.

PUBLICATIONS: *Splash* (quarterly), numerous brochures.

Aquatic Habitat Institute
#180 Richmond Field Station
1301 S. 46th Street
Richmond, CA 94804
(510) 231-9539
(510) 231-9414 (FAX)
Margaret R. Johnston, Executive Director

An independent, nonprofit corporation with a legislative mandate to coordinate all pollution-related research and monitoring programs in the San Francisco Bay-Delta area, AHI is charged with developing the Bay Information Network (BIN). BIN serves as a centralized clearinghouse for information about the estuary and allows easier access to information on current research. AHI has built a portable, computer-based public education display on the Bay and Delta that provides information on the impact of human activities on this delicate area.

PUBLICATIONS: Bibliography of reports, teaching materials, videos, brochures, posters, and a newsletter.

Baltic Marine Environment Protection Commission–Helsinki Commission
Mannerheimintie 12 A
SF-00100 Helsinki
Finland
9-602366
9-602366 (FAX)
Göte Svenson, Chairman of the HELCOM ad hoc high level Task Force

Founded in 1980, this regional organization represents the interests of Denmark, Germany, Poland, Sweden, and Russia. Its purpose is to protect the Baltic Sea area from pollution, especially that produced by industry. It also combats oil and chemical spills. The organization is currently attempting to identify pollution hot spots in the Baltic area and come up with a plan to stop further pollution in the Baltic with the help of the four main international banks in the region.

PUBLICATIONS: *Baltic Sea Environment Proceedings* (more than 40 to date).

Bay Conservation and Development Commission (BCDC)
30 Van Ness Avenue, Suite 2011
San Francisco, CA 94102
(415) 557-3686
(415) 557-3767 (FAX)
Alan Pendleton, Executive Director

BCDC is the state government agency in charge of San Francisco Bay area environmental management. It regulates development in the bay as well as along its shore and undertakes long-range planning for this area.

PUBLICATIONS: *The San Francisco Bay Plan, The Suisun Marsh Protection Plan.*

Bay Institute of San Francisco
10 Liberty Ship Way, #120
Sausalito, CA 94965
(415) 331-2303
(415) 332-8799 (FAX)
David Behar, Executive Director

Founded in 1981, this nonprofit organization researches biological, economic, hydrologic, and legal issues relating to San Francisco Bay. It is active in efforts to protect and restore the Bay-Delta estuary, reform California water policy, and increase fresh water flow to the Bay. The Institute also works to reduce the discharge of selenium and other contaminants into the Bay.

PUBLICATIONS: *Bay on Trial* (quarterly news journal).

Baykeeper
Fort Mason, Building A
San Francisco, CA 94123-1382
(415) 567-4401 or 1-800-KEEPBAY
(415) 567-9715 (FAX)
Michael Herz, Executive Director

Baykeeper is a nonprofit organization founded in 1989 by Michael Herz and modeled on New York's Hudson Riverkeeper program. Baykeeper sponsors volunteers who patrol the San Francisco Bay in their own aircraft and boats to detect violations of environmental law. Baykeeper also monitors water quality, reports pollution incidents, litigates against violators, and receives and acts on citizen complaints. It has six major programs: regulatory agency reform; boatyard toxic paint waste monitoring; coliform bacteria and toxic level monitoring; incident intake, tracking, analysis and follow-up; fish contamination study; and public education.

PUBLICATIONS: *BayKeeper Log* (quarterly newsletter), numerous reports.

California Coastal Commission
45 Fremont, Suite 2000
San Francisco, CA 94105-2219
(415) 904-5200
(415) 904-5400 (FAX)
Adopt-A-Beach: 1-800-COAST-4-U
Peter M. Douglas, Executive Director

This state organization is charged with protecting, maintaining, enhancing, and restoring the coastal environment. It helps coordinate the Adopt-A-Beach program in California and serves as the statewide sponsor of International Coastal Clean-Up day. One of the Commission's programs involves marking storm drains that flow into the sea to raise public awareness (and to help prevent the unsafe disposal of toxic wastes). The commission works closely with the Center for Marine Conservation, and has produced free directories of marine and coastal resources for a number of regions in the state, as well as fact sheets, posters, packets, booklets, letters, stickers, garbage bags, and curriculum guides (grades 2–7).

PUBLICATIONS: Staff reports, educational materials, curriculum guides, books, maps, brochures, posters, legal and research documents, marine and coastal education resource directories.

California Coastal Conservancy
1330 Broadway, Suite 1100
Oakland, CA 94612
(510) 286-1015
(510) 286-0470 (FAX)
Peter Grenell, Executive Director

This state organization was founded to preserve, enhance, and restore coastal resources in California. It helps coordinate the COASTWEEKS and Adopt-A-Beach programs.

PUBLICATIONS: *Coast & Ocean* (quarterly) and numerous brochures.

California State Water Resources Control Board
P.O. Box 100
Sacramento, CA 95812-0100
(916) 657-2390
(916) 657-1258 (FAX)
Walt Pettit, Executive Director

The Board monitors water quality for the entire state through nine regional Water Quality Control Boards. It monitors sources of water pollution, sets water quality standards, and operates monitoring and assessment programs. Its jurisdiction includes groundwater, inland fresh water bodies, bays, estuaries, and ocean waters within three miles of the shore.

PUBLICATIONS: Brochures and numerous internal publications.

Canada–United States Environmental Council (CUSEC)
See **Defenders of Wildlife**

Center for Marine Conservation
1725 DeSales, NW, Suite 500
Washington, DC 20036
(202) 429-5609
(202) 872-0619 (FAX)
Roger E. McManus, President

Founded as the Center for Environmental Education in 1972, this nonprofit membership organization works for the protection of marine wildlife and habitats, and the conservation of coastal and ocean resources. Best known for its leading role in the beach cleanup movement, the CMC also works to prevent the accidental entanglement and drowning of marine animals in fishing gear and debris and to keep plastics out of the marine environment. The organization conducts policy research, promotes public awareness through education, involves citizens in public policy decisions, and supports domestic and international conservation programs for the marine environment. Specific projects have included coordinating the work of thousands of beach cleanup volunteers, and organizing state and industry programs to reduce dangerous marine debris.

PUBLICATIONS: *Marine Conservation News* (quarterly), *Coastal Connection,* (biannual), *A Citizen's Guide to Plastics in the Ocean: More than a litter problem* (monograph), and dozens of books, including two coloring books and 14 specialized information packets.

Center for Oceans Law and Policy
University of Virginia School of Law
Charlottesville, Virginia 22901
(804) 924-7441
(804) 924-7362 (FAX)
John Norton Moore, Director

The Center hosts forums and lectures on ocean law and policy. It is considered an international authority on the Law of the Sea Convention and on maritime law in general, and has one of the best collections of maritime law materials in the world. The Center was founded in 1975.

PUBLICATIONS: *United Nations Convention on the Law of the Sea 1982: A Commentary* (10-volume annotated series that serves as a framework for judicial and governmental interpretation).

Center for Short-Lived Phenomena (CSLP)
P.O. Box 199, Harvard Square Station
Cambridge, MA 02238
(617) 492-3310
(617) 492-3312 (FAX)
Richard Golob, Director

The CSLP was founded in 1968 by the Smithsonian Institution and became a private, for-profit organization in 1975. It specializes in the collection and dissemination of information about oils spills and oil pollution worldwide. The CSLP will undertake information searches, compilations, and investigations on a contractual basis. World Information Systems is the publishing arm of the endeavor.

PUBLICATIONS: *Golob's Oil Pollution Bulletin, Hazardous Materials Intelligence Report.*

Chesapeake Bay Foundation (CBF)
162 Prince George Street
Annapolis, MD 21401
(410) 268-8816
(410) 268-6687 (FAX)
William C. Baker, President

Headquartered in Annapolis, CBF is the largest nonprofit organization (85,000 members) working to save Chesapeake Bay. It works to protect and improve the estuarine environment of the Bay by taking more than 34,000 students a year out in boats to test the quality of the water, fish, dredge for oysters, and provide a sense of what the bay is. CBF runs three island centers and an extensive restoration program and has produced three videos on the bay. It recently won the president's Environment and Conservation Challenge award for its education program.

PUBLICATIONS: *CBF News,* (bimonthly newsletter), annual reports, calendars.

Clean Seas
1180 Eugenia Place, Suite 204
Carpinteria, CA 93013
(805) 684-3838
(805) 684-2650 (FAX)
Darryle Waldron, Manager

This nonprofit, commercial organization was founded in 1970 to provide oil spill response equipment, training, and personnel for the offshore oil industry. It conducts training programs for oil industry personnel, government agencies, and wildlife rehabilitation personnel.

PUBLICATIONS: *Clean Seas Oil Spill Manual,* brochures, and newsletters.

Coast Alliance
235 Pennsylvania Avenue, SE
Washington, DC 20003
(202) 546-9554
(202) 546-9609 (FAX)
Beth Millemann, Executive Director

This private, nonprofit public interest group (founded in 1979) is a national coalition of coastal activists dedicated to raising public awareness about resources on the ocean and Great Lakes coasts. It conducts public education and awareness programs and works on legislation to protect coastal resources. The Alliance organized and ran the Year of the Coast during 1979 and 1980, a public education campaign that helped spur a policy review of federal programs affecting the coasts.

PUBLICATIONS: *And Two If By Sea: Fighting the Attack on America's Coasts* (monograph), *Getting to the Bottom of It: Threats to Human Health and the Environment from Contaminated Underwater Sediments* (report), other reports and documents.

Coastal States Organization
Hall of the States
Suite 322
444 North Capitol Street, NW
Washington, DC 20001
(Telephone and fax not listed, by request).

The CSO was founded in 1970 to serve as the governors' official representative on ocean and coastal affairs of the United States. It does *not* serve as an information center for the public. The CSO advocates sound marine resource management and helps the states participate in the development and implementation of national coastal policy. It guided the rewriting and congressional approval in 1990 of the Coastal Zone Management Act (CZMA), participated in the passage of the Oil Pollution Act of 1990, and served as coordinator and project manager for the National Public Trust Study. The CSO established the CSO Ocean Policy Network, composed of government-appointed representatives, to study and formulate ocean policy. This group challenged the U.S. Army Corps of Engineers' Operation and Maintenance Dredging Regulations.

PUBLICATIONS: Numerous.

CONCAWE
See **Oil Companies' European Organization for Environmental and Health Protection**

Council on Ocean Law
1709 New York Avenue, NW
Washington, DC 20006
(202) 347-3766
(202) 638-0036 (FAX)
Charles Higginson, Executive Director

Founded in 1980 by Elliot Richardson to support the establishment and further development of law for the world's oceans, the COL actively promotes these objectives in international forums. It has an independent panel on the law of ocean uses, prepared papers on the Law of the Sea, and prepared a report for Bill Clinton, before he was elected president, that outlined U.S. interests in the Law of the Sea Convention. The COL sponsors an annual essay contest on an important ocean issue at the Naval War College.

PUBLICATIONS: *Ocean Policy News* (monthly), and Panel on Ocean Law publications (monographs).

Cousteau Society, Inc.
870 Greenbrier Circle, Suite 402
Chesapeake, VA 23320
(804) 523-9335
(804) 523-2747 (FAX)
Jacques-Yves Cousteau, President

This nonprofit, membership-supported organization was founded by Jacques-Yves Cousteau and Jean-Michel Cousteau in 1973 for the protection and improvement of the quality of life for present and future generations. The Society works to protect the integrity of the seas through a series of international expeditions on the Society ships *Calypso* and *Alcyone*, on which valuable film reports are produced. The Society also sponsors scientific research projects and participates in drafting international conventions, treaties, and ocean policy decisions. It is seeking support for a Bill of Rights for Future Generations, a five-point declaration that spells out each generation's responsibility to maintain and protect the heritage of an undamaged planet. The Society hopes the bill will be adopted by the General Assembly of the United Nations.

PUBLICATIONS: *Calypso Log* (bimonthly), *Dolphin Log* (bimonthly), numerous scientific papers.

Earthwatch
P.O. Box 403
680 Mount Auburn Street
Watertown, MA 02272
(617) 926-8200
(617) 926-8532 (FAX)
Brian A. Rosborough, President

This unique organization unites scholars and citizens by sponsoring scientific research expeditions developed by scientists but partly staffed by lay volunteers, who share both the cost of the projects and the work involved. Although only some of Earthwatch's projects have dealt with marine pollution, such projects are becoming increasingly numerous. Earthwatch has mobilized more than 700 research projects in 85 countries, providing researchers with 23,000 volunteers and more than $15 million in funds and equipment.

PUBLICATIONS: *Earthwatch* (monthly), *The Earthcorps Daily Planet* (newsletter).

Earth Island Institute
300 Broadway, Suite 28
San Francisco, CA 94133
(415) 788-3666
(415) 788-7324 (FAX)
John Knox and David Phillips, Executive Directors

Earth Island Institute was founded in 1982 to coordinate environmental and wildlife protection projects. Although the Institute does not focus specifically on the prevention of marine pollution, many of its projects (for example, the International Marine Mammal Project and the Japan Environmental Exchange) do deal directly with this aspect of conservation.

PUBLICATIONS: *Earth Island Journal: An International Environmental News Magazine* (quarterly), *Green Alternative Information for Action* (monthly).

Entanglement Network Coalition
c/o Defenders of Wildlife
1244 19th street NW
Washington, D.C. 20036
(202) 659-9510
(202) 833-3349 (FAX)
Dr. Albert Manville, Director of Science Policy
Chris Croft, Marine Mammal Program Coordinator

The Entanglement Network Coalition (founded in 1983) is a consortium of more than fifty environmental, conservation, and animal protection groups. It attempts to eliminate hazards to marine and freshwater ecosystems resulting from the use of nondegradable materials (such as plastics) at sea. The Coalition reports that more than 750,000 seabirds and 125,000 marine mammals die annually due to entanglement. The Coalition conducts research, prepares studies, and provides speakers.

PUBLICATIONS: *The Entanglement Network Newsletter* (periodic).

Environmental Action, Inc.
1525 New Hampshire Avenue, NW
Washington, DC 20036
(202) 745-4870
(202) 745-4880 (FAX)
Ruth Caplan, Executive Director

This organization works on environmental issues through lobbying, research, education, and organizing, and has tackled issues such as plutonium production, nuclear waste disposal, and glass bottle recycling. It also publicizes the names of those members of Congress who have done the most to *stop* programs designed to help the environment, calling them the "Dirty Dozen." Enviromental Action is not specifically dedicated to marine issues, but has supported land-based actions that do much to help prevent marine pollution.

PUBLICATIONS: *Environmental Action Magazine* (bimonthly).

Environmental Defense Fund (EDF)
257 Park Avenue South
New York, NY 10010
(212) 505-2100
(212) 505-2375 (FAX)
Frederick D. Krupp, Executive Director

Founded in 1971, this nonprofit public interest organization works toward the responsible reform of public policy in the fields of water resources, international environment, and solid waste issues, among many others. EDF initiates litigation, provides public service, and organizes public education campaigns. It sponsors Environmental Information Exchange, which provides access to scientific, legal, regulatory, and economic information. One program is aimed at stopping the thousands of pounds of plastics that are dumped in the ocean each day and on enforcing the Congressional ban on the ocean dumping of plastics. In fact, the EDF was instrumental in generating international agreements in this area. EDF also works to eliminate toxic hazards worldwide—even in Antarctica, where EDF filmed and publicized a large diesel spill. This exposure helped bring about stricter environmental controls in this pristine corner of the globe.

PUBLICATIONS: *EDF Letter* (bimonthly).

Friends of the Earth
218 D Street, SE
Washington, DC 20003
(202) 544-2600
(202) 543-4710 (FAX)
Mrs. Jane Perkins, President

Friends of the Earth was founded in 1969 to preserve biological, cultural, and ethnic diversity and to give people a voice in decisions affecting the environment. It seeks to influence policies and attitudes about issues such as coastal and ocean pollution and corporate accountability. FOE operates the Take Back the Coast project in conjunction with other grassroots organizations. It also provides information on federal and international funding for conservation. FOE identifies critical areas of ocean policy, especially relating to ocean pollution, and sponsors analytical research and policy assessments. It is dedicated to protecting the oceans for the people and wildlife that depend on them for life, livelihood, and enjoyment. The Environmental Policy Institute and the Oceanic Society merged with Friends of the Earth in 1990.

PUBLICATIONS: *Friends of the Earth/Not Man Apart* (ten per year), and numerous pamphlets, handbooks and posters.

The Friends of the Everglades
3744 Stewart Avenue
Coconut Grove, FL 33133
(305) 888-1230
Nancy Carroll Brown, President
Marjory Stoneman Douglas, President Emeritus

This organization grew out of a grassroots effort in 1969 led by Marjory Stoneman Douglas to prevent an airport from being constructed on the Tamiami Trail; the airport would have blocked the flow of fresh water through the Everglades. The Friends of the Everglades continues to work to protect the Okeechobee-Everglades Basin and the Florida Everglades. The Everglades is a unique wetlands habitat of both salt water and fresh water life; it constitutes a rare national treasure and an integral part of the Florida ecosystem. The phone number listed above is a 24-hour hotline for inquiries concerning environmental topics, particularly those dealing with water quality and toxic waste.

PUBLICATIONS: *The Reporter* (sporadic), *Who Knows the Rain* (booklet).

GESAMP
See **Joint Group of Experts on the Scientific Aspects of Marine Pollution**

Get Oil Out, Inc. (GOO)
1114 State Street
Santa Barbara, CA 93101
(805) 965-1519
Henry Feniger, President

GOO was founded in 1969 following an offshore oil platform blowout that dumped thousands of gallons of oil into the Santa Barbara Channel. GOO is one of the oldest grass-roots advocacy organizations specifically dedicated to marine pollution prevention. Its purpose is to monitor and limit (or eliminate) offshore oil and gas operations along the entire California coast. The group has testified at public hearings, and lobbied against and participated in litigation against the oil industry. In 1991, GOO worked with the California Coastal Commission to prevent Chevron from sending tankers through the Santa Barbara Channel. Although it has donated its library to the University of California at Santa Barbara, the organization still serves as an information center on oil pollution and oil development. GOO does not publish.

Greenpeace USA
1436 U Street, NW
Washington, DC 20009
(202) 462-1177
(202) 462-4507 (FAX)
Steve D'Esposito, Acting Executive Director

This well-known, international nonprofit membership organization fights impending ecological disaster in the marine environment while trying to balance the needs of both the fishing industry and marine life. It leads the world's environmental organizations in marine environmental policy formulation by providing environmental education and initiating dramatic "direct actions" (such as intercepting ships transporting hazardous chemicals). It has been very active in the prevention of radioactive marine pollution. Although originally known for its Save the Whales efforts, the organization also works to protect other marine mammals. Greenpeace has also launched a major campaign to force the paper industry to replace chlorine bleaching with oxygen-based bleaching in order to prevent discharge of dioxin into the sea. Greenpeace is now working to help Eastern European countries restore their ravaged ecosystems.

PUBLICATIONS: *Greenpeace Action* fact sheets on hazardous waste, whaling, nuclear testing, greenhouse effect, etc., *Greenpeace* (bimonthly newsletter).

Heal the Bay
1640 5th Street, 112
Santa Monica, CA 90401
1-800-HEAL BAY or (310) 394-4552
(310) 395-6878 (FAX)
Adi Liberman, Executive Director

Heal the Bay, a nonprofit advocacy organization founded in 1985, produces a Beach Pollution Report. This report gives beaches in the Los Angeles area a "grade" based on bacteria levels in the water, and is intended to help beachgoers determine which beaches are safe for swimming.

PUBLICATIONS: *1991 Beach Pollution Report,* 16 pp., *1992 Beach Pollution Report,* 16 pp., and numerous brochures and posters including *20 Ways to Heal the Bay* and *When It Rains, It Pours—Off the Streets and Into the Oceans.*

Hellenic Marine Environment Protection Association (HELMEPA)
5, Pergamou Street
171 21 Nea Smyrni
Athens, Greece
934 3088
935-3847 (FAX)
Dimitris Mitsatsos, Director General

HELMEPA (founded in 1982) is the environmental protection association of Greece. It works to prevent marine pollution both by organizing seminars and publishing books and guidelines aimed at raising the professional standards of shipping personnel. It also wages publicity campaigns to alert and educate the public regarding ways to prevent marine pollution. A three-year program begun in 1991 known as MEDSPA, carried out in conjunction with the European Commission, encompasses both professional training and a beach cleanup effort in the Mediterranean. HELMEPA has also designed software that allows seamen to test their knowledge of the laws on marine pollution prevention. The organization's antipollution mascot is the sea gull.

PUBLICATIONS: *NEA* (monthly newsletter), many guidelines and handbooks for shippers, posters and brochures for the general public.

Inform
381 Park Avenue South
New York, NY 10016
(212) 689-4040
(212) 447-0689 (FAX)
Joanna D. Underwood, President

Inform was founded in 1974 to identify practical ways to protect our natural resources and public health, primarily through waste reduction and management. Research done under its auspices is published in books, abstracts, newsletters, and articles. Staff members speak at congressional briefings and business conferences. Inform conducts case studies of industry practices that affect the environment. It helped shape the first federal legislation on waste reduction and spurred the creation of the EPA's first Office of Pollution Prevention.

PUBLICATIONS: *INFORM Reports* (quarterly), and numerous books including: *A Citizen's Guide to Promoting Toxic Waste Reduction, Garbage Management in Japan,* and *Cutting Chemical Wastes: What 29 Organic Chemical Plants Are Doing To Reduce Hazardous Wastes.*

International Center for the Solution of Environmental Problems
535 Lovett Blvd.
Houston, TX 77006
(713) 527-8711
Dr. Joseph Goldman, Technical Director

The Center is an independent, nonprofit organization founded in 1974. In addition to researching air/water pollutant transfer and other marine pollution issues, the Center also performs investigations and forensic studies for offshore oil projects.

PUBLICATIONS: Quarterly newsletter.

International Marinelife Alliance
94 Station Street, Suite 645
Hingham, MA 02043
(617) 383-1209
(617) 383-0411 (FAX)
Dr. Peter Rubec, President

The Alliance supports practical measures to protect and restore the health and diversity of marine environments worldwide, with emphasis on developing countries. Recent Alliance programs include a study of the causes of the destruction of coral reefs in Costa Rica, the development of a National Marine Disaster Response Plan for the U.S. Congress that will assess and implement cleanup and rehabilitation of marine wildlife harmed by oil spills, and a study of the potential environmental impact of a jetport in Japan on one of the last healthy blue coral reefs off that nation's shores (the jetport project was cancelled).

PUBLICATIONS: *Sea Wind* (quarterly bulletin).

International Maritime Organisation (IMO)
4 Albert Embankment
London SE1 7SR
United Kingdom
Phone: 71-735-7611
71-587-3210 (FAX)
William A. O'Neil, Secretary-General

Established in 1958 and based in London, the IMO was known as the Intergovernmental Maritime Consultative Organisation (IMCO) until 1982. It is the specialized administrative agency within the United Nations for maritime safety, the prevention and control of marine pollution from ships, and related matters. The IMO has more than 130 member nations but no direct powers of enforcement. It promotes the safety of international shipping and prevents marine pollution caused by ships. It also works to prevent land-based pollution that has an impact on the ocean, such as dredging operations and the transport by sea of pollutants and waste generated on land. The IMO prepares international standards and regulations for safe shipping and and promotes their implementation. It encourages and assists in the establishment and operation of regional and national contingency plans for responding to major incidents involving oil and other harmful substances. The IMO has produced a series of conventions designed to deal with illegal discharges of oil.

PUBLICATIONS: *IMO News* (quarterly, in English), *Focus on IMO* (series of documents dealing with the work of IMO, including *Preventing Marine Pollution, Chemicals at Sea, Dumping at Sea,* and many others).

International Ocean Disposal Symposium
See **International Ocean Pollution Symposium**

The International Oceanographic Foundation (IOF)
4600 Rickenbacker Causeway
P.O. Box 499900
Miami, FL 33149-9900
(305) 361-4888
(305) 361-4711 (FAX)
Victoria Myers, Executive Director

Founded in 1953, this membership organization provides authoritative, unbiased information about the oceans. It encourages both scientific investigation and development of the sea. The IOF runs a Q&A service called Sea Secrets, offers scholarships for graduate level study in marine science, and sponsors an educational travel program for members. The "IOF Gold Medal" is an award for nonscientists who have advanced the scientific study of the ocean. The Sea Frontiers Mail Order Department offers books and gifts.

PUBLICATIONS: *Sea Frontiers* (bimonthly), *Training & Careers in Marine Science* (information packet).

International Ocean Pollution Symposium
Department of Oceanography, Ocean Engineering &
Environmental Science
Florida Institute of Technology
Melbourne, FL 32901
(407) 768-8000 ext. 8008
(407) 984-8461 (FAX)
Dr. Iver Dudall, Director

The Symposium, founded in 1978 as the International Ocean Disposal Symposium, promotes the exchange of ideas and information among investigators researching the problems associated with waste disposal in the ocean. Following exchanges that take place primarily at annual meetings, attendees draft recommendations and guidelines for scientific studies of current and future ocean waste disposal practices. Symposium presentations are documented in peer-reviewed books. The group maintains a library with more than 3,000 documents concerning ocean waste disposal.

PUBLICATIONS: *Marine Pollution Bulletin* (annual), *Wastes in the Ocean* (monographs series), *Oceanic Processes in Marine Pollution* (monograph series), *International Ocean Pollution Symposium* (monograph series).

MEETINGS: International Ocean Disposal Symposium (annual).

International Tanker Owners Pollution Federation Ltd. (ITOPF)
Staple Hall, Stonehouse Court
87-90 Houndsditch
London EC3A 7AX
United Kingdom
Phone: 71-621-1255, or 42-691-4112 (after office hours)
71-621-1783 (FAX)
Dr. I. C. White, Managing Director

ITOPF is a nonprofit organization founded in 1968 by the oil companies and independent tanker owners. ITOPF's 3,500 members operate 6,500 tankers, with a total capacity of approximately 156 million gross tons. This represents nearly 97 percent of the the world's tanker tonnage.

ITOPF administers the Tanker Owners Voluntary Agreement concerning Liability for Oil Pollution (TOVALOP). The group responds to marine oil spills, analyzes claims for compensation, conducts contingency planning, and provides advisory services and training. ITOPF has extensive experience in oil spill cleanup and damage assessment.

PUBLICATIONS: *Response to Marine Oil Spills* (video/film/book), *Ocean Orbit* (annual newsletter), numerous booklets.

Joint Group of Experts on the Scientific Aspects of Marine Pollution (GESAMP)
c/o International Maritime Organization
4 Albert Embankment
London SE1 7SR
United Kingdom
Phone: 1 7357611
Administrative Secretary

Founded in 1969, this "organization of organizations" is a multidisciplinary body of independent experts. The International Maritime Organization (IMO) provides the administrative staff for this U.N. body. The sponsoring U.N. agencies include: IMO, FAO, UNESCO, WMO, WHO, IAEA, and UNEP. More than 350 scientists from at least 50 nations have participated in GESAMP Working Groups. GESAMP provides scientific advice on pollution and other marine and coastal environmental problems. It recently has adopted the concept of "sustainable development," which implies that "the present use of the environment and its resources shall not prejudice the use and enjoyment of that environment and its resources by future generations." The Joint Secretariat of GESAMP establishes the subjects to be considered by GESAMP during its annual five-day working sessions. Presentation of Working Group Reports is the main activity at these sessions.

PUBLICATIONS: *GESAMP Reports and Studies* (irregular).

Keep Britain Tidy Group
See **Tidy Britain Group**

Louisiana Universities Marine Consortium
8124 Highway 56
Chauvin, LA 70344
(504) 851-2800
(504) 851-2874 (FAX)
Dr. Paul Sammarco, Executive Director

The Louisiana Universities Marine Consortium is a state-run, nonprofit organization of 13 public universities, 4 private universities, and several state agencies in Louisiana founded in 1979 to help coordinate marine research and education by providing excellent coastal field and laboratory facilities for these programs. The Consortium operates a marine center with laboratories, several research vessels, and satellite laboratory facilities. It offers college courses in marine sciences emphasizing field experience and the study of living organisms, as well as public education programs. The Consortium studies the environmental effects of oil, gas, and petrochemical industries (particularly offshore oil development in the Gulf of Mexico) and the impact of nutrient enrichment on the ocean. A National Shellfish Pollution Indicator study is also underway.

PUBLICATIONS: Annual report and numerous articles in professional journals.

Marine Debris Information Office (MDIO)
See **Center for Marine Conservation**

Marine Forum for Environmental Issues
c/o Department of Zoology
The Natural History Museum
Cromwell Road
London SW7 5BD
United Kingdom
Phone: 71-938-9114
71-938-9158 (FAX)
Mrs. Swantje-A. Brodie Cooper, Administrator

Founded as the North Sea Forum in 1987, the Marine Forum today is an umbrella organization concerned with environmental issues in all coastal shelf areas of northwest Europe and the North Atlantic region. Its purpose is to improve communication on marine coastal issues, to follow up on recommendations and implementation, to centralize and coordinate responses of participants to major marine and coastal issues, and to promote the aims of the World Conservation Strategy for the sustainable use and development of the marine and coastal environment. It holds four plenary meetings each year.

PUBLICATIONS: Annual reports. *The Marine Forum 1990 North Sea Report* is currently available.

MEETINGS: Four plenary meetings each year.

Marine Pollution Control Unit (of the DTp) (MPCU)
Room 6/1
Sunley House
90/93 High Holborn
London WC1V 6LP
United Kingdom
Phone: 71-405 6911 ext. 3153
Fax: 71-831 7681/71-405 0794
Mr. C. J. Harris, M.B.E., Head of the Unit

MPCU operates under the Department of Transport. It was founded in 1979 to coordinate and implement the UK government's response to oil or chemical pollution incidents in territorial waters. It maintains contingency plans and has resources needed for marine cleanup operations, including specially equipped aircraft. MPCU also provides advice and assistance to local authorities (who are primarily responsible for coastal cleanup), maintains stockpiles of specialized beach cleanup equipment, and funds a research program that studies both onshore and ocean cleanup methods. MPCU compiles and processes government claims for compensation from polluters and their insurers, and follows up on reports of possible illegal discharges of oil at sea.

PUBLICATIONS: Primarily informational brochures and annual reports.

Marine Science Institute
500 Discovery Parkway
Redwood City, CA 94063-4715
(415) 364-2760
(415) 364-0416 (FAX)
Robert E. Rutherford, President and founder

Founded in 1970 to increase environmental awareness and provide marine science education, MSI works primarily through schools and colleges. It runs three main programs, the best known of which is the Discovery Voyage (for fifth-grade students and older), in which teachers use a packet of materials provided by the Institute to prepare students for a four-hour trip onto the bay to study plankton, sedimentation, and marine wildlife. The Shoreside program provides a similar experience onshore. The Institute has an aquarium mounted on a trailer (called the Inland Voyage) which is taken directly to the schools. More than 200,000 school children have participated in these hands-on programs.

PUBLICATIONS: Instructional materials.

National Audubon Society
950 3rd Avenue
New York, New York 10022
(212) 979-3000
National Legislative Hotline: (202) 547-9017
Peter Berle, President

This powerful national organization (founded in 1905) may be best known for organizing bird walks. However, it also organizes beach cleanups and supports an extensive staff of wardens to patrol wildlife sanctuaries. In the minds of many it is synonymous with conservation and ecology.

PUBLICATIONS: *Audubon Activist* (bimonthly), among others.

National Coalition Against the Misuse of Pesticides
701 E Street, Suite 200
Washington, D C 20005
(202) 543-5450
Jay Feldman, National Coordinator

Founded in 1981, the Coalition focuses public attention on serious pesticide poisoning problems and promotes reduced pesticide exposure through alternative pest management strategies that use few or no toxic chemicals. Although the group's initial concern was pesticide poisoning of food, its efforts are equally valuable in protecting the ocean from airborne and riverborne pesticide contamination. The group has taken on the U.S. Environmental Protection Agency's policies in this area and has established a broad coalition of health, environmental, labor, farm, consumer, and church groups.

PUBLICATIONS: *Pesticides And You* (newsletter, 5 issues annually), brochures, booklets, compilations and a monthly technical report.

National Coalition for Marine Conservation
5105 Paulsen Street, Suite 243
Savannah, GA 31405
(912) 354-0441
(912) 354-0234 (FAX)
Ken Hinman, Executive Director

Founded in 1973 to promote the conservation of marine resources and the protection of the ocean environment, NCMC seeks legislative and administrative solutions to problems threatening marine resources. It has dealt with ocean dumping issues and habitat problems.

PUBLICATIONS: *Marine Bulletin* (bimonthly newsletter, now includes *Currents*), *Ocean View* (sporadic), *Proceedings of the Marine Fisheries Symposium* (annual).

National Ocean Industries Association
1120 G Street NW, Suite 900
Washington, DC 20005
(202) 347-6900
(202) 347-8650 (FAX)
Franki Stuntz, Director of Administration & Member Services

This nonprofit association, founded in 1972, lobbies to promote the business interests of marine industries by encouraging "environmentally sound development and use of ocean resources." It supports legislation favorable to marine industries and seeks to expand the role of the free enterprise system in the development of ocean resources.

PUBLICATIONS: *Membership Directory* (annual), *Washington Report* (bimonthly), brochures.

National Oceanographic and Atmospheric Administration (NOAA)
U.S. Department of Commerce
Office of Public Affairs
Washington, DC 20230
(202) 482-2985
(202) 482-3154 (FAX)
John A. Knauss, Undersecretary for Oceans & Atmosphere

Prior to the establishment of NOAA in October 1970 under the aegis of the Department of Commerce, ocean-related functions existed in many different government agencies. Even after these agencies were combined to form NOAA, most retained their independent nature and continued their work on the many facets of marine pollution. The agencies include the National Ocean Service at (202) 606-4003; the National Marine Fisheries Service at (301) 713-2370; the Hazardous Material Response Branch at (206) 526-6046; and the National Environmental Satellite, and Data and Information Service (NESDIS) at (301) 763-7190. NOAA also funds many independent marine pollution-related efforts, such as the Marine Debris Information Center (*see* Center for Marine Conservation) and the Sea Grant Program (*see* Sea Grant Program). Through these various programs, NOAA helps the states manage their coastal zones, monitors water quality, establishes marine and estuarine sanctuaries, operates satellite monitoring systems, and conducts research of all kinds on the oceans and atmosphere. NOAA also maintains an extensive marine pollution library collection (301) 443-8330.

PUBLICATIONS: Numerous.

National Science Foundation (NSF)
Division of Ocean Sciences
1800 G Street NW
Washington, DC 20550
(202) 357-9639
(202) 357-7621 (FAX)
Grant Gross, Director of Division of Ocean Sciences

The NSF provides funding for oceanographic research, including research on marine pollution prevention. It is the source of as much as one-third of the funding for programs at the nation's oceanographic research institutes and universities, and also awards numerous fellowships. Globec is an NSF special program in marine biosciences; its projects deal with global change, especially alterations in the marine environment.

National Toxics Campaign (NTC)
1168 Commonwealth Ave.
Boston, MA 02134
(617) 232-0327
John O'Connor, Executive Director

Founded in 1984, NTC unites citizens to prevent pesticide pollution, and reduce toxic wastes. NTC has called for a nationwide moratorium on new toxic waste incinerators and dumps until companies are required by federal regulation to reduce their use and discharge of toxic substances. It fights to prohibit the use of 70 cancer-causing chemicals and the sale of these chemicals to Third World nations. It has opened the Citizen's Environmental Laboratory at Boston University to provide low-cost, reliable testing for 56,000 pesticides, petrochemicals, and other pollutants. NTC has accused the U.S. Environmental Protection Agency of improperly advocating toxic waste incineration. Although the group fights toxic chemical contamination wherever it occurs, its activities help protect the marine environment.

PUBLICATIONS: *Toxic Times* (quarterly newsletter).

Natural Resources Defense Council (NRDC)
40 W. 20th Street
New York, NY 10011
(212) 727-2700
(212) 727-1773 (FAX)
John H. Adams, Director

NRDC is a nonprofit environmental advocacy organization founded in 1970 and dedicated to the wise management of natural resources through research, public education, and the development of environmental policy. Its concerns include coastal protection, air and water pollution, nuclear safety and energy production, toxic substances, and the protection of wilderness and wildlife. NRDC works to increase public understanding of how law may be used to protect natural resources, engages in litigation that may set widely applicable precedents or preserve natural resources, and monitors federal departments and regulatory agencies concerned with the environment. It opened the door to mutual nuclear test ban verification. NRDC also issues periodic beach reports that grade the safety of U.S. costal swimming areas.

PUBLICATIONS: *Amicus Journal* (quarterly), *NRDC Newsline* (bimonthly), *Truly Loving Care: For Our Kids and For Our Planet* (quarterly newsletter), and numerous brochures, pamphlets, books, and reports.

Netherlands Institute for the Law of the Sea (NILOS)
Janskerkhof 3
3512 BK Utrecht
The Netherlands
31-30-39 3060
31-30-39 3073 (FAX)
Prof. Mr. A. H. A. Soons, Director

Established in 1984 as part of the Faculty of Law of the University of Utrecht, NILOS now operates within the Netherlands Institute for Social and Economic Legal Research (NISER) of the Utrecht Law Facility. NILOS conducts research on all issues related to the law of the sea, with special emphasis on questions relating to the conservation, utilization, and management of marine natural resources. It assists states (especially developing nations) in dealing with law of the sea issues. It covers boundary and fishery issues, as well as the legal aspects of marine pollution. Research and training projects are the principal work of NILOS, resulting in papers and publications presented at international conferences by NILOS staff members on various problems of the law of the sea.

PUBLICATIONS: *International Organizations and the Law of the Sea Documentary Yearbook* (annual), *NILOS Newsletter* (biannual), among others.

North Sea Directorate
P.O. Box 5807
2280 HV Rijswijk
The Netherlands
31-703-94-95-00
31-703-90-06-91 (FAX)
Dik Tromp, Director

Founded in 1971 under the auspices of the Water Management Department of the Ministry of Transport, Public Works, and Water Management, the North Sea Directorate manages the waters and ocean floor of the Dutch portion of the North Sea. It patrols the sea to monitor and prevent pollution, collects and distributes ecological information, conducts aerial surveillance, issues dumping and dredging licenses, and develops policies regarding national and international use of the sea. The Directorate owns a specially equipped vessel, the *Smal Agt,* that can be sent to the scene of an oil spill.

PUBLICATIONS: *North Sea: Action and Policy, North Sea Action Plan, Combatting Oil Pollution in the North Sea, The North Sea: A Sea of Opportunity or the Sewer of Europe?,* among other informational brochures and reports.

Nuclear Information and Resource Service (NIRS)
1424 16th Street, N.W., Suite 601
Washington, D.C. 20036
(202) 328-0002
(202) 462-2183 (FAX)
Michael Mariotte, Director

An advocacy organization founded in 1978, NIRS is dedicated to a sound nonnuclear energy policy. In the area of marine pollution, the focus is on preventing the dumping of nuclear wastes in the ocean and other waterways, and on educating the public regarding the environmental hazards of nuclear energy, such as the problems that result from flushing radioactive coolant water into the ocean.

PUBLICATIONS: *Nuclear Monitor* (semimonthly).

OCA/PAC
See **United Nations Environment Programme (UNEP)**

Ocean Pollution Research Center
Rosenstiel School of Marine and Atmospheric Science
University of Miami
4600 Rickenbacker Causeway
Miami, Fl 33149
(305) 361-4160
(305) 361-4701 (FAX)
Dr. Christopher Mooers, Director

The Ocean Pollution Research Center was formed in June 1992, following the *Exxon Valdez* oil spill. This incident focused attention on the fact that South Florida, the Caribbean, and the Gulf of Mexico are also vulnerable to oil spills due to heavy tanker traffic and cruise ship operations, as well as Cuba's recent offshore oil exploration activities. The Rosenstiel School of Marine and Atmospheric Science is one of the largest oceanic and atmospheric science graduate schools in the southeastern United States. The Center takes a "generic approach" to ocean pollution, with an initial focus on oil trajectory (how oil disperses)and fate modelling (what eventually happens to the oil spilled) in the Straits of Florida and the Gulf of Mexico, including regional oceanic and atmospheric circulation models and observation systems (i.e., Ocean Surface Current Radar). The Center develops and demonstrates new methodologies potentially useful in handling oil spills, and also does oil spill impact studies.

PUBLICATIONS: Under development.

The Oceanic Society
See **Friends of the Earth**

Oil Companies' European Organization for Environmental and Health Protection (CONCAWE)
Madouplein 1
B-1030 Brussels
Belgium
32-2-220 31 11
32-2-219 46 46 (FAX)
Denis Lyons, Director

CONCAWE was founded in 1963 as the European oil refining companies' international study group for clean air and water. The acronym CONCAWE is derived from CONservation of Clean Air and Water in Western Europe. Originally focused on studying oil spill cleanup technology, CONCAWE's role has expanded to include the task of limiting the effects of oil pollution after a spill has occurred. CONCAWE serves in an advisory capacity during oil spill cleanups.

PUBLICATIONS: *Oil Spill Dispersant Efficiency Testing; Methods of Prevention, Detection and Control of Spillages in West European Oil Pipelines; Characteristics of Petroleum and its Behavior at Sea;* and *Strategies for the Assessment of the Biological Impacts of Large Coastal Oil Spills—European Coasts,* among numerous others.

Oil Companies International Marine Forum (OCIMF)
15th Floor
96 Victoria Street
London SW1E 5JW
United Kingdom
Phone: 71 828-7696/6283
71 245 2921 (FAX)
Mr. E. J. M. Ball, Director

OCIMF is a voluntary association of oil companies with an interest in the marine transport and storage of crude oil, gas, and petrochemicals. OCIMF was formed by 18 companies at a meeting held in London in April 1970. By 1991, membership included 35 companies and groups worldwide. OCIMF promotes safety and the prevention of pollution from tankers and at terminals. Formation of the OCIMF was the oil industry's initial response to increasing public awareness of oil pollution after the *Torrey Canyon* spill. OCIMF has consultative status with IMO, sponsors research activities, and prepares and publishes procedural guidelines.

PUBLICATIONS: More than 40 booklets and manuals, many procedural in nature, and many co-sponsored with other organizations.

Paris Commission
New Court
48 Carey Street
London WC2A 2JE
United Kingdom
Phone: 71 242-9927
71 831-7427 (FAX)

The Paris Commission was founded in 1974 to uphold the Convention for the Prevention of Marine Pollution from Land-based Sources. It sets regulations and standards, and conducts research.

PUBLICATIONS: *Annual Report* (annual, French and English), *The First Decade* (book, in French and English).

Reef Relief
P.O. Box 430
Key West, FL 33041
(305) 294-3100
(305) 293-9515 (FAX)
Bruce Etshman, President

Reef Relief is an environmental group battling threats to a band of coral reefs that stretches 166 miles, from the tip of Florida toward Cuba. The reefs lie near a busy shipping lane and have been damaged by boat groundings and pollution. The group has produced a rap song called "Don't Teach Your Trash To Swim" for use as a public service announcement.

PUBLICATIONS: *Reef Line* (quarterly), Pamphlets: *Florida's Coral Reef Ecosystem, Household Guide to Coral Reef Protection, Handbook of Environmentally Safe Business Practices for the Hospitality Industry.*

Regional Seas Activity Center and Regional Seas Research Studies
See **United Nations Environment Programme (UNEP)**

Resources for the Future, Inc.
Quality of the Environment Division
1616 P Street, NW
Washington, DC 20036
(202) 328-5000
(202) 265-8069 (FAX)
Robert W. Fri, President

This is an independent, nonprofit research and educational organization founded in 1952 to study the environmental consequences of human activity and how public decisions affecting the environment are made. Past research projects include estimates of the recreation benefits from water pollution control, the development of alternatives for reducing the generation of hazardous wastes, and an assessment of natural resource damage from oil spills and releases of hazardous substances. Resources for the Future provides congressional testimony, prepares articles for national newspapers, and sponsors workshops and seminars.

PUBLICATIONS: Technical Discussion Papers, ranging in price from $2.25 to $5.00 and including titles such as *Economics and Nutrient Reductions in the Chesapeake Bay, Innovative Policies for Sustainable Development in the 1990s: Economic Incentives for Environmental Protection, "Black Mayonnaise" and Marine Recreation: Methodological Issues in Valuing a Cleanup.*

Save the Bay
434 Smith Street
Providence, RI 02908
(401) 272-3540, 849-8430
(401) 273-7153 (FAX)
Curt Spalding, Executive Director

Founded in 1970 when a small group of neighbors banded together to take on a powerful oil company planning to build a refinery in Tiverton, Rhode Island, this membership organization works to clean up polluted rivers and ocean bays, reopen closed shellfish beds, and bring failing sewage treatment plants up to environmentally safe standards—in general, to protect the Narragansett Bay ecosystem. The group is even developing a satellite tracking system to help avoid oil spills. It was active in the cleanup following the 300,000 gallon *World Prodigy* oil spill at the mouth of Narragansett Bay and has also worked closely with the state's Environmental Quality Study Commission.

PUBLICATIONS: *Save the Bay* (bimonthly newsletter), *Save the Bay Junior Newsletter, Annual Report.*

Scripps Institution of Oceanography
University of California, San Diego
Mail Code 0233
La Jolla, CA 92093
(619) 534-3624 (Public Affairs)
(619) 534-5306 (FAX)
Dr. Edward A. Frieman, Director

The George H. Scripps Memorial Marine Biological Laboratory was founded in 1903 and integrated into the University of California in 1906. Scripps is a research center and a graduate training facility. It also provides public services in the marine sciences, including the Stephen Birch Aquarium-Museum, which disseminates information on research underway at the Institution and helps educate the public on wise use of our oceans and ocean resources. The Institution is open to the general public. It does basic research in biology, geology, chemistry, physical oceanography, and the like, needed by other scientists who carry out applied research on marine pollution.

PUBLICATIONS: *Preparing for a Career in Oceanography* (8-page booklet, $2.00), *Bulletin of the Scripps Institution of Oceanography, Annual Report.*

The Sea Grant College Program
NOAA Sea Grant, R/OR1
SSMC-1
1335 East-West Highway
Silver Spring, MD 20910

(Phone and fax not listed, by request. For further information, call the Sea Grant Depository, University of Rhode Island (401) 792-6114)
David B. Duane, Director

The National Sea Grant College program was established by the U.S. Congress in 1966, as the first federal program to support activity across the full spectrum of the marine sciences. It was conceived by Senator Claiborne Pell and other legislators as a unique plan to develop and use wisely our nation's marine and Great Lakes resources through university-based research, education and training, and technology transfer. The Sea Grant program is based at more than 30 universities in coastal and Great Lake states, with ties to more than 300 universities and research organizations nationally. Sea grant programs foster greater understanding, wise use, and conservation of the marine environment. Funding is provided primarily by the National Oceanographic and Atmospheric Administration (NOAA) and is matched at each institution. Annually the National Sea Grant Office issues priority guidance to the network of Sea Grant programs. Program directors meet several times each year to discuss these priorities and submit proposals.

PUBLICATIONS: *Sea Grant Abstracts* (quarterly).

Seas At Risk
Vossiusstraat 20-lll
NL-1071 AD
Amsterdam
The Netherlands
31-20-675-4336
31-20-675-3806 (FAX)

Seas At Risk was founded in 1986 to improve the exchange of information on marine pollution issues among national and international organizations. It promotes discussion on international policy-making and legislation and coordinates international activities aimed at increasing public awareness of marine pollution. It also operates SEAPRESS, a press-information bureau, publishes press releases on marine environment developments, and provides the media with background information.

PUBLICATIONS: *North Sea Monitor* (quarterly).

Sierra Club
Sierra Club Clean Coastal Waters Task Force
1841 N. Fuller Ave., 209
Los Angeles, CA 90046
(213) 874-6732

This subgroup of the Sierra Club focuses on the marine environment. The Clean Coastal Waters Task Force was formed to ensure that all sewage receives at least secondary treatment, to eliminate ocean sludge dumping, and to restore the quality of our coastal waters. The task force focuses on the southern California coast.

PUBLICATIONS: *Waste Watchers* (bimonthly newsletter). Sierra Club national publications: *Sierra* (bimonthly), *Sierra Club National News Report* (26 issues per year). Also publishes books on environmental issues.

Skidaway Institute of Oceanography
P.O. Box 13687
McWharter Drive
Savannah, GA 31416
(912) 598-2453
(912) 598-2751 (FAX)
Dr. David Minzel, Director

The Skidaway Institute is a small research facility that does important original research on both organic and inorganic pollutants, including petroleum products, chlorophenols, and trace metals. It also develops instrumentation to help produce analytical data.

PUBLICATIONS: *Georgia Marine Science Center Technical Report Series* (irregular).

Smithsonian Institution
Washington, DC 20560
(202) 357-2627
(202) 786-2377 (FAX)
Dr. Robert McCormick Adams, Secretary

This prestigious organization was founded in 1846 to preserve and articulate our cultural and natural heritage. That heritage now includes the marine environment. The Smithsonian Environmental Research Center in Edgewater, MD, near the Chesapeake Bay, studies land-water relationships and plans public programs to increase awareness of ecological systems and how they are affected by human activities. The Smithsonian Institution Marine Station at Link Port, in Fort Pierce, FL, studies estuarine and marine environments along Florida's eastern coastline and the adjacent ocean shelves.

PUBLICATIONS: *Smithsonian Magazine* (monthly), and numerous books.

Student Environmental Action Coalition
P.O. Box 1168
Chapel Hill, NC 27514
(919) 967-4600
(919) 967-4648 (FAX)
Beth Ising, National Office Representative

This organization was founded in 1988 by students working for environmental justice. Although not focused exclusively on marine environmental issues, the Coalition does serve as a clearinghouse on environmental issues. It encompasses 17 regional groups and 2,000 campus and high school organizations, allowing it to coordinate national and regional activities. A substantial number of the individual organizations are involved in marine pollution issues.

PUBLICATIONS: *Threshold* (8 per year newsletter), *Campus Ecology* (An environmental audit of campuses and textbook), *The Student Environmental Action Guide, High School Organizing Guide, Student Organizing Guide.*

Tidy Britain Group
The Pier
Wigan, Lancashire WN3 4EX
United Kingdom
Phone: 0942-824620
0942-824778 (FAX)
Barbara Sinker, Executive Director

Founded in 1953 as Keep Britain Tidy, this membership organization was concerned primarily with land litter and beautification efforts until the mid-1970s. It was renamed Tidy Britain Group in 1988, and is now the "government's national agency for litter abatement and environmental improvement." In 1977 it conducted the first survey of marine litter in the UK ("Discarded Containers on a Kent Beach"). The group prepares annual Marine Litter Research Reports and launched the Tidy Beach Awards program in 1987.

PUBLICATIONS: *Marine Litter Research Reports* 1–6, an extensive annual report, and numerous brochures.

TOVALOP
See **ITOPF,** International Tanker Owners Pollution Federation Ltd.

United Nations Environment Programme (UNEP)
P.O. Box 0552
Nairobi, Kenya
2 333 930
Dr. Mostafa K. Tolba, Executive Director

UNEP was established by the United Nations General Assembly in 1972 as a coordinating body to monitor and plan environmental activities, to promote environmental cooperation on the regional level, and to organize intergovernmental conferences on the protection of the environment. UNEP environmental programs having to do with the world's oceans include the Global Environmental Monitoring System and the

Regional Seas Activity Center, which is part of the Oceans and Coastal Programme Activity Center. Under the auspices of this program, UNEP has facilitated the establishment of 11 regional marine environmental programs. UNEP monitors changes in the environment via EarthWatch.

PUBLICATIONS: Numerous.

United Nations Educational, Scientific, and Cultural Organization (UNESCO)
7, Place de Fontenoy
F-75700, Paris
France
1-456-81000
Frederico Mayor, Director General

UNESCO includes several organizations concerned with the marine environment, including the International Oceanographic Commission (IOC), which coordinates oceanographic programs globally. Among these programs is the Marine Pollution Research and Monitoring program, which assesses oceanic contaminants and the biological effects of pollution. Another is the Working Committee for the Global Investigation of Pollution in the Marine Environment.

PUBLICATIONS: Numerous.

United States Coast Guard
Norfolk Federal Building
Norfolk, VA 23510
(804) 441-3307
Commanding Officer

Although originally set up for search and rescue and to enforce safety at sea, during the early 1970s the enforcement of a series of marine environment protection regulations fell under the aegis of the U.S. Coast Guard. Coast Guard vigilance is today one of the only means of enforcing such laws as the Water Quality Improvement Act, the Federal Water Pollution Control Act amendments, and the Ocean Dumping Act, among others. The Coast Guard responds to pollution complaints, enforces federal regulations, and inspects and documents commercial vessels.

United States Department of Energy
1000 Independence Avenue, SW
Washington, DC 20585
(202) 586-5000
(202) 586-5049 (FAX)

In addition to enforcing safety and operating regulations for offshore oil and gas installations and for some oil transporters, the Department of Energy (DOE) is also charged with regulating the transport, storage and/or disposal of hazardous and radioactive wastes.

United States Environmental Protection Agency (US EPA)
Office of Wetlands, Oceans and Watersheds
401 M Street, SW
Washington, DC 20460
(202) 260-7751 (Public Information)
(202) 260-6294 (FAX)
Bob Wayland, Director

The EPA is the federal agency charged with protecting the marine and estuarine environment, and with general environmental regulation and enforcement. This includes issuing disposal and dumping permits, and monitoring and undertaking research projects. Each regional office of the EPA handles special marine pollution projects, most notably Regions 1 through 7, which encompass coastal areas. The EPA studies hazardous wastes, radioactivity, pesticides and other environmental pollutants in terms of risk assessment, disposal, and impact. More than one-quarter of the agency's budget is directed toward water issues, including marine pollution.

PUBLICATIONS: Numerous.

United States Geological Survey
Water Resources and Geologic Division
12201 Sunrise Valley Drive
Reston, VA 22092
(703) 648-5215
Phil Cohen, Chief Hydrologist
Dallas Peck, Geologic Division Director

The Water Resources Division of the USGS measures both the quantity and quality of all the water in the nation. It is a research projects organization; it has no regulatory responsibility. It conducts several research on marine pollution, including studies of contaminants in coastal areas, as well as the transport of contaminants from rivers and estuaries into coastal waters. It also manages USGS natural resources surveys in response to the Superfund Act of 1980. The Geologic Division of the USGS studies areas of the ocean within 200 miles of the U.S. coast that have been contaminated. This research focuses on ocean sediments and the transport of contaminants.

PUBLICATIONS: Technical reports.

Warren Spring Laboratory (WSL)
Gunnels Wood Road
Stevenage
Hertforshire SG1 2BX
United Kingdom
438 74-1122
438 36-0858 (FAX)
Dr. Doug Cormack, Director and Chief Executive

Established in 1959, this lab is the UK government's primary source of environmental research. It succeeded the former Fuel Research Station, and became an executive agency of the Department of Trade and Industry in 1989. WSL helps achieve environmentally sustainable development by providing government and industry with cost-effective ways to comply with environmental standards. It offers technical services and consulting in monitoring environmental quality and pollution abatement. WSL is best known for its development of oil cleanup devices (e.g., Springsweep), its role in the Earth Resources Satellite Program (aerial surveillance), and the development of EUROSPILL, a PC-based simulation to predict the spread of oil spills.

PUBLICATIONS: Dozens of monographs, consisting primarily of research reports of a highly technical nature (catalog available).

The Wilderness Society
900 Seventeenth Street, NW
Washington, DC 20006-2596
(202) 833-2300
(202) 429-3958 (FAX)
George Frampton, President

Founded in 1935, the Wilderness Society is the only national conservation organization devoted primarily to public lands protection and management issues. In this pursuit, it also protects important marine resources, such as the Florida Everglades. It employs a combination of advocacy, analysis, and public education in its campaigns to improve the management of America's national parks, forests, wildlife refuges, coastal areas, and Bureau of Land Management lands. At the top of its agenda are the prevention of oil drilling in the Arctic National Wildlife Refuge and the restoration of the Florida Everglades ecosystem.

PUBLICATIONS: *Wilderness* (quarterly), *Annual Report,* numerous reports.

Woods Hole Oceanographic Institution
Coastal Research Center
Woods Hole, MA 02543
(508) 548-1400, Ext. 2900
(508) 457-2172 (FAX)
Robert Beardsley, Director

The Coastal Research Center at Woods Hole is an interdisciplinary marine science group founded in 1979 to focus on coastal issues. Its areas of interest include the effects of waste disposal on coastal ecology, effects of oil spills, contamination transport, ecotoxicology, and the assimilative capacity of the coastal ocean.

PUBLICATIONS: Technical reports.

Woods Hole Oceanographic Institution
Marine Policy Center
Woods Hole, MA 02543
(508) 548-1400, Ext. 2449
(508) 457-2184 (FAX)
Director, James Broadus

The Marine Policy Center at Woods Hole, founded in 1972, brings law, economics, management, and other social sciences to bear on marine environmental issues, through postdoctoral fellowships and various programs for visiting investigators and the general public. Woods Hole Oceanographic Institution itself was founded in 1888 as the Biological Station at Woods Hole, and is one of the most influential marine research institutes in the world.

PUBLICATIONS: Research results in journals, especially *Oceanus* (published quarterly by Woods Hole Oceanographic Institution).

World Environment Center
419 Park Avenue South, Suite 1800
New York, NY 10016
(212) 683-4700
(212) 683-5053 (FAX)
Antony G. Marcil, President & CEO

The World Environment Center was founded in 1974 with the support of the United Nations Environment Programme (UNEP) to foster cooperation in international environmental management and industrial safety. It places pollution control experts in eligible developing countries.

PUBLICATIONS: *Pollution Prevention Pays* (brochure), *Chlorine Manual,* newsletter.

Libraries

Arthur J. Morris Law Library
University of Virginia
Charlottesville, VA 22901
(804) 924-3384

One of the most extensive and complete collections on ocean law in the United States.

National Oceanographic and Atmospheric Administration Library
Department of Commerce
6009 Executive Blvd.
Rockville, MD 20852
(301) 443-8330 (reference)
Has an extensive collection on marine pollution issues. Will not undertake research on behalf of the general public, but the collection is open to the public and the materials are available via interlibrary loan.

National Sea Grant Depository
Pell Library Building
University of Rhode Island
Bay Campus
Narragansett, RI 02882
(401) 792-6114

As the official Sea Grant Depository library, this collection gives a comprehensive look at the documents produced by all of the Sea Grant programs since their inception. A maximum of ten documents at a time may be borrowed free of charge for one-month periods. Because it houses the Depository and because it serves the University of Rhode Island's research needs, the Pell Library is also a very valuable source of marine pollution materials.

6

Selected Print Resources

IN RESEARCHING THE FIRST EDITION of *Marine Pollution: Diagnosis and Therapy,* Professor Dr. Sebastian Gerlach, noted marine scientist at the Institut für Meereskunde (in the former Federal Republic of Germany), noted that in 1971 he had only to review "some one hundred bibliographic items." Just four years later, "no fewer than 868 publications under the heading of 'Marine Pollution' were reported." During the two decades since he made this observation, the quantity of materials in this field has increased by at least an order of magnitude.

This chapter contains only the key print materials that contribute in some unique way to an understanding of the field. Reference materials, including yearbooks, bibliographies, directories, atlases, encyclopedias, and handbooks, are listed first. Only journals dealing primarily with marine pollution issues are included, although pertinent articles do also appear in general environmental publications. (Please note that many of the publications by the institutes and organizations listed in Chapter 4 also contain valuable information on the marine environment.) A few important government publications are then listed. Books and articles make up the majority of this chapter. Following a general listing where materials covering a broad range of marine pollutants are cited, there is a specialized subject listing in which materials are divided into sections by type of pollution as in Chapter 1: Sewage, Marine Debris, Toxic Chemicals, Heavy Metals, Oil, and Radioactive Materials. Every entry is annotated.

Reference Materials

ACOPS Year Book 1986–1987. Advisory Committee on Protection of the Sea (ACOPS). Oxford/New York: Pergamon Press, 1987. 120 pp.

Global in scope, this books contains the names, addresses, and phone numbers of hundreds of organizations involved in activities having to do with the ocean. It is a comprehensive list and one of the only sources for non–U.S. agencies.

Barnett, Judith B. **Marine Science Journals and Serials: An Analytical Guide.** (Annotated Bibliography of Serials: No. 7). Westport, CT: Greenwood Press, 1986. 191 pp.

Although this valuable bibliography covers a far broader scope than just pollution in the marine environment, it is annotated and can be searched by subject. A valuable tool.

Brown, Lester, et al. eds. **State of the World: A Worldwatch Institute Report on Progress toward a Sustainable Society.** New York: W. W. Norton & Co., 1984–.

This annual publication provides an excellent overview of the environmental condition of the planet, including the marine environment.

California Coastal Commission. **Marine and Coastal Educational Resources Directory.** San Francisco, CA: California Coastal Commission, 45 Fremont, Suite 2000, San Francisco, CA 94105-2219. November, 1990. 2 vol.

This photocopied directory is divided into sections for northern and southern California. It covers a broad range of environmental organizations active in the state, and also includes reference materials and speakers available for school and community organizations. Free on request.

Center for Marine Conservation. **Marine Debris Educational Materials.** Washington, DC: Center for Marine Conservation. August, 1991. 21 pp.

This unbound, photocopied directory is the best source for information regarding posters, public service announcements, videos, slide shows, advertisements, logos, pamphlets, booklets, brochures, fact sheets, curriculum guides, coloring books, stickers—even imprinted garbage bags—having to do with marine debris, entanglement, and beach cleanup efforts. Many of the materials identified are free and all are designed for use in schools, by clubs, or by the general public. The directory is organized by state (and country), and provides complete ordering information. Free on request.

Champ, Michael A. and P. Kilho Park. **Global Marine Pollution Bibliography: Ocean Dumping of Municipal and Industrial Wastes.** New York: IFI/Plenum Press, 1982. 399 pp.

This was one of the first comprehensive bibliographies on this critical source of marine pollution at the time it was published. Much has since been written on the topic, but this remains a landmark document.

Christol, Carl Q. **Ocean Pollution in the Marine Environment: A Legal Bibliography.** Washington: U.S. GPO, 1971. 93 pp.

This document was the first publication of the Sea Grant program (Sea Grant Publication No. 1). It was published in December 1971, with support from Grant GH-89 of the National Sea Grant program, as well as by the U.S. Department of Commerce with the University of Southern California (1971). It was one of the first compilations of its kind.

Corson, Walter H., ed. (The Global Tomorrow Coalition). **The Global Ecology Handbook: What You Can Do about the Environmental Crisis.** Boston: Beacon Press, 1990. 414 pp.

Chapter 8, Ocean and Coastal Resources, contains valuable data and statistics on marine pollution issues. It is focused on the U.S., with an emphasis on the pollution of coastal waters.

Couper, Alastair, ed. **The Times Atlas and Encyclopedia of the Sea.** New York: HarperCollins Pubs., Inc., 1989. 272 pp.

Well written, well organized, and full of valuable graphics on the scientific, legal, economic, and environmental aspects of the ocean. Numerous articles present shipborne commerce statistics, pollution sources and statistics, coastal data, sea law, and much more in a two-page-spread format. The most useful chapter for marine pollution data is entitled *The World Ocean.* Up-to-date information in an easy-to-use format. Includes 11 appendices, a glossary, a bibliography, and an index.

Environment Abstracts. New York: R. R. Bowker. Monthly, cumulated annually.

Covers all facets of the environment, not just pollution. Use subject heading "marine pollution." Provides excellent access to primarily scholarly and technical materials for those who wish to consult the original studies that are cited in the more popular literature.

Hanson, M. Bradley and Patricio Aguilar, eds. **Marine Debris Bibliography.** Seattle: Center for Quantitative Sciences, University of Washington, April 1989. 71 pp.

This valuable bibliography, which is global in scope and actually includes several bibliographies bound together, provides nearly 2,000 complete bibliographic references for books, articles, and documents concerning marine debris. It is divided into a very general first section which contains a wide variety of references, followed by sections on beach surveys, marine plastics, net entanglement, and tar balls. Available through the Center for Marine Conservation, 1725 DeSales NW, Ste. 500, Washington, DC 20036, (202) 429-5609.

International Environmental Reporter. Washington, DC: BNA Books (Division of the Bureau of National Affairs), Weekly, 1978–.

Covers all facets of pollution. Use subject headings "marine pollution" and the *see also* references listed therein. This is an excellent source of succinct and accurate information about events (such as tanker spills) and international agreements. The index/summary breaks topics down geographically by nation or region.

Kormondy, Edward J., ed. **International Handbook of Pollution Control.** Westport, CT: Greenwood Press, 1989. 466 pp.

This comprehensive reference book highlights pollution problems and controls in all the major nations of the world, providing an excellent source of background information on a nation's marine pollution activities. Specific marine pollution information is provided for Australia, Italy, Korea, Mexico, Singapore, Taiwan, and the United Kingdom. Marine pollution legislation information is also provided for numerous other nations.

Miller, E. Willard and Ruby M. Miller. **Environmental Hazards: Water Pollution, a Bibliography.** Monticello, IL: Vance Bibliographies, 1985. 49 pp.

This excellent bibliography was published in January 1985 as part of the Public Administration Series. Only a portion of it pertains to marine pollution, but it documents a number of difficult to identify publications.

National Oceanographic and Atmospheric Administration. **Directory of the NOAA Library and Information Network.** Rockville, MD: NOAA Assessment and Information Services Center, Department of Commerce, 6009 Executive Blvd., Rockville, MD 20852. Updated periodically.

Provides data on the holdings of more than 35 NOAA information centers and libraries around the nation. Many of these collections contain materials on marine pollution.

Netherlands Institute for the Law of the Sea (NILOS). **International Organizations and the Law of the Sea Documentary Yearbook.** The Netherlands: Martinus Nijhoff Publishers, 1991. 800 pp.

An annual publication that reprints the most important documents concerning the law of the sea issued each year by international organizations.

Pfafflin, J. R. and E. N. Ziegler, eds. **Encyclopedia of Environmental Science and Engineering.** New York: Gordon Breach, 1983. 3 vols.

Contains readable definitions of marine pollution terms and concepts.

Reducing World Pollution: A Compilation of United Nations and U.S. Government Documentary Materials. (Taylor's Encyclopedia of Government Officials, Federal and State. Debate packet, 1992–93). Dallas, TX: Political Research, Inc., 1992. 61 p.

The scope of this useful reference book is broader than marine pollution, but it includes references to documents on marine environmental policy.

Seredich, John, ed. **Your Resource Guide to Environmental Organizations.** Irvine, CA: Smiling Dolphins Press, 1991. 514 pp.

Includes the purposes, programs, accomplishments, volunteer opportunities, publications, and membership benefits of 150 environmental organizations around the world. About 20 percent of these organizations are directly or indirectly concerned with marine pollution prevention. At least as valuable as the organizational data provided is the name index and the biographical sketches of fourteen environmental leaders.

Trzyna, Thaddeus C. and Roberta Childers. **World Directory of Environmental Organizations: A Handbook of National and International Organizations and Programs.** 4th edition. Sacramento, CA: California Institute of Public Affairs, 1973–.

This comprehensive global handbook includes more than 2,100 national and international organizations and programs (governmental and non-governmental) concerned with protecting the earth's resources in more than 200 countries. The scope is far broader than marine pollution, but it includes many organizations around the world that work in this area.

Wang, James C. F. **Ocean Politics and Law: An Annotated Bibliography.** (Bibliographies and indexes in law and political science, 16). Westport, CT: Greenwood Press, Inc., 1991. 243 pp.

This selective bibliographic survey of recent literature on ocean and maritime law, legal practice, and related issues from the 1960s to mid-1980s includes a handy list of basic sources. It also has more detailed listings on new technological advances and applications, environmental protection, control of pollution (including oil pollution), natural resource exploitation (from offshore oil to deep seabed mining), oceanography, regulation of commerce, and broader issues of international law. The more than 2,000 sources cited range from general encyclopedias and popular articles to specialized journals, monographs, and yearbooks. Some entries are annotated.

Journals

Coastal Connection
Center for Marine Conservation
1725 DeSales N.W., Suite 500
Washington, DC 20036
Biannual.

This newsletter describes beach cleanup activities around the country.

Golob's Oil Pollution Bulletin
World Information Systems
P.O. Box 199, Harvard Square Station
Cambridge, MA 02238
Biweekly.

This important newsletter is edited by Richard Golob and published by World Information Systems, the publishing arm of the Center for Short-Lived Phenomena (CSLP). Covers pollution prevention, control, and cleanup worldwide. Provides some of the best data available on oil spills, much of it generated internally by the Center.

Hazardous Materials Intelligence Report
World Information Systems
P.O. Box 199, Harvard Square Station
Cambridge, MA 02238
Weekly.

This international newsletter focuses on the safe management of hazardous materials and hazardous wastes, including transportation, storage, disposal, and emergency response. Provides timely information on new regulations and legislation, hazardous materials spills and hazardous waste site contaminations, damage claims, litigation, insurance settlements, cleanup equipment, and technological breakthroughs.

Marine Conservation News
Center for Marine Conservation
1725 DeSales N.W., Suite 500
Washington, DC 20036
Quarterly.

The official newsletter of the Center for Marine Conservation. Devoted exclusively to marine debris and entanglement topics. Written in clear, concise language appropriate for students and nonscientists.

Marine Pollution Bulletin
Pergamon Press, Inc.
660 White Plains Rd.
Tarrytown, NY 10591
Monthly.

Contains news and articles about marine pollution around the world. Some of the articles are technical in content, but most are written for a nontechnical audience. Indexed by Engineering Index and Nuclear Science Abstracts. Began in June 1968 as the North East Pollution Programme Bulletin.

Oceanus
Woods Hole Oceanographic Institution
Woods Hole, MA 02543
Quarterly.

"Translates" the scientific and highly technical work being done at the Institution into lay terms. An extremely valuable resource in the field of marine pollution.

Tide
Coastal Conservation Association
4801 Woodway, Suite 220 W
Houston, TX 77056
Bimonthly

Focuses on conservation issues and activities from Texas to Florida and South Carolina, with information on educational programs, legislation, and legal issues.

Government Publications

Bioremediation for Marine Oil Spills. Washington, DC: Congress of the U.S., Office of Technology Assessment, G.P.O., 1991. 31 pp.

Background paper on the use of bacteria and other biological resources to clean up oil spills. This document describes the process and the procedure, without evaluating any possible adverse effects of this methodology.

Commission on Marine Science, Engineering and Resources. **Our Nation and the Sea.** Washington, DC: Government Printing Office, 1969. 4 vol.

This landmark document marked the beginning of an intense national focus on the oceans, primarily as a resource to be exploited. It recommended, among other things, the creation of a national ocean agency. Out of this recommendation was born the National Oceanographic and Atmospheric Administration (NOAA). Makes little or no mention of protecting the marine environment.

Cottingham, David. **Persistent Marine Debris: Challenge and Response: The Federal Perspective.** Washington, DC: NOAA, 1988. 41 pp.

Addresses the bioaccumulation of toxic chemicals in the food chain and suggests regulatory steps.

Federal Council for Science and Technology. **Oceanography—The Ten Years Ahead.** Interagency Committee on Oceanography (ICO) pamphlet No. 10. Washington, DC: Government Printing Office, June 1963.

A statement of the U.S. government's long-range goals and how they would be accomplished. There is no mention of concern regarding the possible impact of these activities on the health of the oceans.

GESAMP. **The State of the Marine Environment.** (IMO/FAO/UNESCO/WMO/WHO/IAEA/UN/UNEP Joint Group of Experts on the Scientific Aspects of Marine Pollution). Boston: Blackwell Scientific Publications, 1991. 128 pp.

Updates the 1972 GESAMP publication which describes the Group's work. Declares that the global ocean is in no peril, but that some coastal waters are in danger. Includes bibliographical references and index.

Interagency Committee on Ocean Pollution Research, Development and Monitoring. **National Marine Pollution Plan, 1981–1985.** Washington, DC: Department of Commerce, April 1981.

Addresses marine pollution issues, primarily in terms of protecting human health and fisheries without hindering the exploitation of marine resources.

Managing Troubled Waters: the Role of Marine Environmental Monitoring. Assessment of Marine Environmental Monitoring, Marine Board, Commission on Engineering and Technical Systems, National Research Council. Washington, DC: National Academy Press, 1990. 125 pp.

Defines the role of environmental monitoring in the management of marine pollution. Provides up-to-date information on the technology now available for marine environmental monitoring, including satellite resources. Water quality monitoring stations, as well as the U.S. laws and legislation that seek to control marine pollution, are reviewed. The text does not, however, indicate what is done when a problem is detected.

Mielke, James E. **Effects of Man's Activities on the Marine Environment.** Washington: GPO, 1975. 135 pp.

Prepared at the request of senators Warren G. Magnuson, chairperson, Committee on Commerce, and Ernest F. Hollings, chairperson, National Ocean Policy Study.

National Academy of Sciences. **Oceanography 1960–1970.** Washington, DC: National Academy of Sciences, 1961, 1962, 1963.

These reports identified national marine policy goals, suggested how they could be met, and proposed a set of milestones for the decade ahead. Senator Pell called them "the first authoritative view of the great potential of the oceans."

National Academy of Sciences. **Radioactivity in the Marine Environment.** Washington, DC: National Academy of Sciences, 1971. 262 pp.

This report, prepared by the Panel on Radioactivity in the Marine Environment of the National Research Council's Committee on Oceanography, provides a detailed account of knowledge gained about radionuclides in the ocean between 1957 and 1970. Although highly technical, this was a landmark work in an area where little information had been previously available.

National Academy of Sciences-National Research Council. **The Effects of Atomic Radiation on Oceanography and Fisheries.** (NAS-NRC Publication No. 551). Washington, DC: National Academy of Sciences, National Research Council, 1957. 137 pp.

One of the first appraisals of radioactivity in the marine environment, published at a time when fallout from the detonation of nuclear devices was the principle source of man-made radionuclides in the oceans. Summarizes the knowledge and problems known at that time and makes suggestions for future research programs. Of historical importance.

National Academy of Sciences, National Research Council. **Radioactive Waste Disposal into Atlantic and Gulf Coastal Waters.** Washington, DC: National Academy of Sciences, National Research Council, 1959.

Prepared at the request of the Atomic Energy Commission, this document accurately reflects scientific opinion of the 1950s regarding radioactive waste disposal. It suggests 28 carefully selected sites, each at least 75 miles from the next, where it was maintained that low-level radioactive wastes could be safely dumped into the ocean. When the public learned of this report, there was outrage. This may have been one of the incidents that triggered the environmental movement in the United States during the 1960s.

National Advisory Council on Oceans and Atmosphere. **National Ocean Goals and Objectives for the 1980s.** In *Ocean Services for the Nation,* January 1980.

A very telling picture of U.S. ocean policy during the decade of the 1980s, one which emphasizes use and exploitation without a clear environmental perspective.

National Oceanic and Atmospheric Administration (NOAA). **Federal Plan for Ocean Pollution Research, Development and Monitoring.** Washington, DC: U.S. Department of Commerce, NOAA. August 1979.

This first plan covers fiscal years 1979 through 1983. Of historical relevance.

National Research Council. Panel on Radioactivity in the Marine Environment. **Radioactivity in the Marine Environment.** Washington: National Academy of Sciences, 1971. 272 pp.

Prepared in response to a request by the Atomic Energy Commission as a comprehensive review of the understanding of the marine environment at that time.

United Nations. **Law of the Sea: Protection and Preservation of the Marine Environment.** New York: United Nations, 1990. 95 pp.

A collection of the texts of international agreements relating to sections 5 and 6 of Part XII of the UN Convention on the Law of the Sea.

United States Department of Commerce. **U.S. Ocean Policy in the 1970s: Status and Issues.** Washington, DC: U.S. Department of Commerce, October 1978. 51 pp.

A readable, albeit bureaucratic, review of the U.S. position on the ocean. Focuses primarily on exploitation, rather than on protection, of the marine environment.

United States National Advisory Committee on Oceans and Atmosphere. **The Role of the Ocean in a Waste Management Strategy: A Special Report to the President and Congress.** Washington, DC: The Committee, GPO, 1981. 104 pp.

Applies the scientific principles of dilution and dispersion, and economic justifications, to recommend dumping of waste matter.

United States National Research Council. **Oil in the Sea: Inputs, Fates and Effects.** Ocean Sciences Board, Commission on Physical Sciences, Mathematics and Resources, National Research Council. Washington, DC: National Academy Press, 1985. 601 pp.

This report estimates that 21 million barrels of oil enter the ocean each year—many times more than the 600,000-plus barrels that were spilled on average each year during the early 1980s. Street runoff, ships flushing their tanks, and effluent from industry account for much of this burden.

Books and Articles

General

Baines, John D. **Protecting the Oceans.** (Conserving Our World Series). Austin, TX: Raintree Steck Vaughn Library, 1990. 48 pp.

Discusses the importance of the oceans, as well as the sources and effects of marine pollution and misuse. Final chapter looks at ways to protect the sea. Good overview written in a very easy-to-read style.

Bright, Michael. **The Dying Sea.** Illustrated by Ron Hayward. New York: Gloucester Press, 1988. 32 pp.

This well-written book appeals to a wide age range and provides a good overview of the current condition of the ocean and the many ways in which humans are damaging this vast resource. Marine dumping, oil and chemical pollution, and their effects on marine creatures paint a bleak but accurate picture of the state of the ocean.

British National Committee on Oceanic Research, Marine Pollution Sub-Committee. **The Effects of Marine Pollution: Some Research Needs.** London: Royal Society, 1979. 78 pp.

This document was originally a memorandum. It is of historical interest, providing an inside look at what were considered priority research topics in Europe at the close of the 1970s.

Brownlee, Shannon. **"Stopping Coastline Pollution at the Sewer and at the Farm."** *U.S. News & World Report* 107 (August 21, 1989): 52.

This article goes beyond the visible debris on Atlantic beaches and documents the deeper problem, including 21 billion gallons of toxic chemicals and raw sewage that the United States deposits in the ocean every day. Very well written and full of important data.

Bulloch, David K. **The Wasted Ocean.** New York: Lyons & Burford, 1989. 150 pp.

This short, summary description of the ocean environment stands out because of its clarity, authority, and explicitness, placing it among the best available books about the marine environment for the general reader. Basics such as population increases and movements to the coast, flooding, waste disposal, farming, fishing, and sources of smog and acid rain are discussed. Then a systematic tour of problems along the U.S. coast (metals in Saco Bay, Maine; Boston's sewage system failures; loss of oysters in Chesapeake Bay; tributyltin in Norfolk Harbor; and California's closure of clam beds) provides striking examples of marine pollution. The book closes with concrete suggestions for action.

Capper, John, Garrett Power, and Frank R. Shivers, Jr. **Chesapeake Waters: Pollution, Public Health and Public Opinion, 1607–1972.** Centreville, MD: Tidewater Publishers, 1983. 217 pp.

This difficult-to-find document is an excellent historical overview of pollution in Chesapeake Bay and public perceptions regarding this extremely delicate and critical U.S. estuary. A 6-page bibliography is included.

Chabreck, Robert H. **Coastal Marshes: Ecology and Wildlife Management.** Minneapolis, MN: University of Minnesota Press, 1988. 138 pp.

A good overview of marsh habitats, with particular emphasis on the Gulf of Mexico coastline. Environmental influences on marshlands, with some interesting examples of marsh management plans that have had unexpected outcomes, are discussed at length. This book shows how human activities can easily destabilize and damage these crucial coastal areas. Well illustrated, with a good glossary of scientific terms. An excellent first book on marsh ecology.

Clark, Robert B. **Marine Pollution.** 2nd rev. ed. New York: Oxford University Press, 1989. 240 pp.

This book was first published in 1986, and a second, revised edition was published in 1989. It is based on the author's "introductory course of lectures on marine pollution" and deliberately tries to avoid value judgements. The first edition took the point of view that waste disposal in the ocean is necessary and simply must be done properly, while the later edition recognizes a need to reduce the volume and types of wastes discharged into the ocean. Clark also describes the high costs of high environmental standards and suggests that a balance be struck. Eight types of pollution are described, the North, Mediterranean, Baltic, Caribbean and Caspian seas are looked at individually, pollution assessment is addressed, and an excellent "Further Reading" bibliography is provided. Widely used as a textbook.

Cousteau, Jacques Yves. **The Sea in Danger.** New York: The Danbury Press, 1973. 144 pp. This is volume 19 of the series **The Ocean World of Jacques Cousteau.**

With color photographs on nearly every page, this is an excellent introduction to the full spectrum of issues in the area of marine pollution, from the balance of nature, to the politics of marine pollution prevention, to pollutants that originate in the air.

D'Elia, Christopher F. **"Nutrient Enrichment of the Chesapeake Bay."** Environment 29 (March, 1987): inside cover, 2–3, 6–11+.

D'Elia is one of the principal scientific figures involved in efforts to understand and save the Chesapeake. This is a technical article that explores the eutrophication controversy, and also contains a sidebar with a legislative history of the Bay.

Denis, Corinne. **"The Sea Has Its Limits."** *World Press Review* 36 (June 1989): 55.

This article was excerpted from *L'Express* of Paris. It downplays the impact of radioactive pollution, emphasizes the coastal impact of toxic wastes in the sea, and explains the Eros 2000 Program, an umbrella organization of the European Community that is striving to understand the ocean and prevent coastal pollution. Provides interesting insights on the European perspective, in English.

Ebb Tide for Pollution: Actions for Cleaning Up Coastal Waters. New York: Natural Resources Defense Council, 1989. 42 pp.

This extremely readable summary of pollution problems addresses the impact of marine pollution on coastal areas and sources of marine pollution. Chapter Three outlines a dozen ways in which marine pollution can be prevented, and the following section details the Natural Resources Defense Council plan to restore coastal waters. The book ends with suggestion on pollution prevention tactics at home, as a consumer, and as a citizen. Sarah Chasis was instrumental in producing this nicely illustrated color booklet, and the foreword is by Jean-Michel Cousteau.

Ehrlich, Paul. **"Ecocatastrophe!"** Ramparts Magazine, 8 (September, 1969) pp. 24–28.

Ehrlich, a biology professor when he wrote this article, later became a very well known environmentalist. This imaginary scenario, the centerpiece of which is the death of the ocean towards the end of the summer of 1979, struck a chord with the American public, and aided the lay person's comprehension of the ocean as part of a very delicately balanced ecosystem. The same article was also published under the title **"Our Plundered Planet: The Death of the Oceans,"** in *Currents* 111 (October, 1969) pp. 23–32.

Fine, John C. **Oceans in Peril.** New York: Atheneum, 1987. 128 pp.

Eight chapters cover the basics of plant and animal resources in the sea, the impact of human activities on these resources, as well as possible solutions to the ever growing problem of marine pollution. Black and white photographs by the author enhance this extremely well-prepared book. One of the best and most complete books on the ocean for students.

Friedman, Wolfgang. **The Future of the Oceans.** Woodstock, NY: Beekman Publications, 1971. 132 pp.

Although this book is not specifically dedicated to marine pollution, it provides an extremely comprehensive, global look at the attitudes (legal, economic, and so on) toward the sea that underlie man's willingness or reluctance to pollute the marine environment.

Gerlach, Sebastian A. **Marine Pollution: Diagnosis and Therapy.** Berlin/New York: Springer-Verlag, 1981. 218 pp.

Originally published in German in 1976, this book offers a European oceanographer's valuable perspective on the problem of marine pollution. Pollution is divided into the categories domestic effluents, industrial effluents, pollution by ships, oil pollution, radioactivity, heavy metals, and chlorinated hydrocarbons; also included is a chapter on the reasons these substances are so harmful to the sea (toxicity, accumulation, and geochemical processes). Includes a section on laws preventing marine pollution and a 16-page bibliography.

Getting to the Bottom of It: Threats to Human Health and the Environment from Contaminated Underwater Sediments. Report. Washington, DC: Coast Alliance, April, 1992.

A clear look at a relatively new concern in the field of marine pollution. Since many ocean contaminants settle to the bottom, additional long-term health impacts of ocean pollution may originate in ocean sediments.

Goldberg, Edward D. **The Health of the Oceans.** Paris: The UNESCO Press, 1976. 172 pp.

One of the first books to speculate on the future health of the oceans. Served as the initial basis for the work of the then newly formed UNESCO Working Committee for the Global Investigation of Pollution in the Marine Environment. This book is particularly valuable because of the author's concern regarding the lack of sufficiently sensitive measurement devices and his willingness to speculate on hazards that might be lurking just beyond science's ability to detect them. The text was reviewed by a number of the author's scientific colleagues and is technical in tone and texture. Supersedes Goldberg's **A Guide to Marine Pollution** published in 1972.

Goldsmith, Edward, et. al. **Imperiled Planet: Restoring Our Endangered Ecosystems.** Cambridge, MA: The MIT Press, 1992. 288 pp.

This reprint of a 1990 publication is an exquisite volume that examines every corner of the earth in chapters organized by geographic feature. The introductory section entitled *Gaia: The Living Planet* is mandatory reading, as are the three chapters dealing specifically with the marine environment: *Wetlands and Mangroves, Coasts and Estuaries,* and *Seas and Oceans.* All are beautifully written and contain a plethora of easily understood information on the sea, how all of its geological features and inhabitants are interrelated, and the impact of human activity on its health. Color photographs from around the world, graphs and maps, and poignant quotes illustrate this comprehensive text.

Goodavage, Maria. **"Murky Waters."** *Modern Maturity* 32 (August-September, 1989): pp. 44–50.

Gives an historical overview of our perception of the ocean, a good rundown on the many pollutants currently poisoning the seas, an explanation of the impact of the ozone hole on the state of the ocean, and a brief look at global efforts to protect the marine environment, including a list of some U.S. organizations involved in this effort. One of the key articles on marine pollution.

Gourlay, K. A. **Poisoners of the Seas.** London and Atlantic Highlands, New Jersey: Zed Books Ltd., 1988. 256 pp.

As the title indicates, this book minces no words regarding the irreparable harm being inflicted upon the seas by mankind. It severely indicts scientists and scientific organizations around the world (GESAMP, in particular) as mealymouthed and euphemistic, and systematically presents solid facts—dates, times, places, amounts, impacts—to reveal the extent of the problem of marine pollution. Part Two is divided into four chapters that correspond to four of the main sources of marine pollution: oil, sewage, hazardous chemicals and heavy metals, and radioactivity. The state of all major seas is addressed, as is the impact of pollution in Third World areas and out-of-the-way zones like the Antarctic. Mr. Gourlay presents a dark view of the future in Part Three. An excellent bibliography is included.

Heyerdahl, Thor. **"How To Kill an Ocean."** *Saturday Review* 3 (November 29, 1975): pp. 12–18.

This eyewitness account of pollution in some of the remotest reaches of our oceans, combined with knowledgeable, thoughtful, and evenhanded warnings as to the foolhardiness of our continuing pollution of the planet's greatest resource, may have done as much to further the general public's awareness of the problem as any effort to date. Written in clear, persuasive language.

Hood, Donald W. **Impingement of Man on the Oceans.** New York, Toronto: Wiley-Interscience, 1971. 713 pp.

A well-thought out and reasoned book, with lots of scientific information presented in readable terms. Sections cover the ways in which pollution enters the sea, different types of marine pollution, the coastal environment, and marine policy and law. The data is somewhat out of date, but the conclusions remain valid.

Horton, Tom and William Eichbaum. **Turning the Tide.** Washington, DC: Island Press, 1991. 250 pp.

The best overview available of North America's largest estuary. Read this in conjunction with the video: **Chesapeake—Living Off the Land.**

Johnston, R. **Marine Pollution.** London: Academic Press, 1976. 703 pp.

Technical, but quite well written. Although dated, this landmark monograph is still accurate and very comprehensive.

Liptak, Karen. **Saving Our Wetlands and Their Wildlife.** New York: Watts, 1991. 63 pp.

Describes the various types of wetlands and wetland wildlife, and clearly outlines the environmental threats to them. A list of wetlands to visit in the United States and the organizations concerned with protecting them is included. A bit overly optimistic and simplistic concerning the future of our wetlands, but a good, clear overview nonetheless.

Marx, Wesley. **The Frail Ocean: A Blueprint for Change in the 1990s and Beyond.** Old Saybrook, CT: Globe-Pequot Press, 1991. 224 pp.

This excellent volume brings Marx's 1967 book, **The Frail Ocean,** up to date, and draws upon the author's many years of research and writing on the sea to recommend courses of action that must be taken to preserve the sea. Written in nontechnical, even lyrical, language. The finest of many fine books on the ocean by Marx.

———. **The Oceans, Our Last Resource.** San Francisco, CA: Sierra Club Books, 1981. 332 pp.

A well-researched, readable book. Although only Chapter 21 (*The Dump of Last Resort*) is specifically on marine pollution, there are references throughout to activities that generate pollution and jeopardize the integrity of the sea.

———. **The Protected Ocean: How To Keep the Seas Alive.** New York: Coward, McCann and Geoghegan, 1972. 94 pp.

One of Marx's earliest books on the sea. It is ahead of its time and hopeful in tone.

Millemann, Beth. **And Two If by Sea: Fighting the Attack on America's Coasts: a citizen's guide to the Coastal Zone Management Act and other coastal laws.** Washington, DC: Coast Alliance, 1986. 109 pp.

An excellent guide to the Coastal Zone Management Act, published by the Coast Alliance as part of its public education and advocacy work. Examines offshore oil and gas development, ocean dumping, and coastal water pollution, among other topics.

Moorcraft, Colin. **Must the Seas Die?** Boston: Gambit, 1973. 194 pp.

A powerfully written exposé on the impact of marine pollution on marine life. Served as a primer of sorts on marine pollution for the growing environmental movement. Although the statistics are out of date, the book remains almost as accurate a look at the threat to the oceans in the 1990s as it was in the 1970s.

"Our Threatened Planet." *Maclean's* 101 (September 5, 1988): 2, 38–43+.

This in-depth cover story does an excellent job at taking a look at a wide range of ecological problems and issues, including overpopulation, the ozone hole, and the Biosphere II experiment. It warns of large-scale global problems unless we dramatically change our ways—soon. This issue of *Maclean's* also includes a second article entitled **"Warning from the Seas,"** by Ann Findlayson (pp. 48–49).

Robinson, W. Wright. **Incredible Facts about the Ocean, Vol. 3: How We Use It, How We Abuse It.** (Ocean World Library). New York: MacMillan Child Group, 1990. 128 pp.

This large, illustrated volume is very basic in tone, but provides valuable information in an easy-to-read format. The other volumes of the series deal primarily with the physics of the ocean.

Sedge, Michael H. **Commercialization of the Oceans.** New York: Franklin Watts, 1987. 128 pp.

Chapter eight of this well-written, easy-to-read book covers the techniques of incineration at sea fairly and in some detail. This process involves the transport of toxic wastes on specially designed incineration ships, and the subsequent incineration of these wastes at sea. This book also includes a brief review of the ocean dumping of sewage sludge by New York and New Jersey, and the dumping of dredged sludge by the Army Corps of Engineers. The first seven chapters of this book focus on ways to use the ocean profitably.

Simon, Anne W. **Neptune's Revenge: The Ocean of Tomorrow.** New York: Franklin Watts, 1984. 222 pp.

This beautifully written volume is an eloquent litany of marine abuses. It is a plea to respect and preserve the vast sea of life. Includes an excellent, if outdated, 24-page bibliography.

Stone, Roger D. **The Voyage of the *Sanderling*.** New York: Knopf Press, 1990. 302 pp.

Compares current environmental conditions along the Atlantic coastline of the Americas with those described by the earliest explorers of the New World. To research the book, Stone sailed from Maine to Rio de Janeiro over a two-year period. The book explores coastal ecology and pollution issues through firsthand accounts of pollution. Contains useful maps and pictures.

Tesar, Jenny. **Threatened Oceans.** (Our Fragile Planet Series). New York: Facts On File, 1992. 128 pp.

Clearly explains the delicate nature of the planet's ecosystem and the ocean's role in that system. **Threatenend Oceans** details the severe environmental burden that humans are currently placing on the oceans, and the many adverse effects of human activities, with special emphasis on ocean dumping. Includes many photographs, both color and black and white.

Tomasevich, J. **International Agreements on Conservation of Marine Resources, with Special Reference to the North Pacific.** Stanford, CA: Stanford University Food Research Institute, 1943. 297 pp.

Contains maps, tables, and diagrams on the conservation of marine resources, primarily fisheries. Also has the full text of many of the first marine pollution agreements ever made.

Thorne-Miller, Boyce and John G. Catena. **The Living Ocean: Understanding and Protecting Marine Diversity.** (Island Press critical issues series, No. 4). Washington, DC: Island Press, 1991. 180 pp.

This book calls attention to the great diversity of life in the marine environment and the need for its protection from pollution, habitat destruction, and overexploitation. It presents well-written discussions on the various marine ecosystems and includes a chapter on international conventions and programs. The final chapter, devoted to future needs, presents many useful guidelines. Could be used as a textbook for environmental issues courses or for general courses in marine biology. Includes a glossary of scientific terms.

Wang, James C. **Handbook on Ocean Politics and Law.** Westport, CT: Greenwood Press, Inc., 1992. 592 pp.

This clear, readable, and very accessible handbook on the law of the sea begins with a general discussion of the oceans and the regulatory challenges they pose. It then explores developments in key sectors, including living and nonliving resources, marine pollution, navigation, and scientific research. It includes analyses of all of the major multilateral agreements relating to the law of the sea and highlights important bilateral and regional initiatives. Contains more than 75 maps and tables, 6 lengthy appendices, and an extensive bibliography.

Whipple, A. B. C.. **Restless Oceans.** (Planet Earth Series). Alexandria, VA: Time-Life Books, 1983. 176 pp.

In the fine tradition of Time-Life Books, **Restless Oceans** does an excellent job covering the topic. Although ocean currents is the primary subject matter, marine pollution is dealt with extensively.

World Commission on Environment and Development. **Our Common Future.** London: Oxford University Press, 1987. 400 pp.

This is actually the report of an International Commission chaired by Norwegian prime minister and internationally known environmentalist Gro Harlem Brundtland (*See* Chapter Three). Although marine pollution is but one facet of this must-read book, it clearly explains the biosphere's burdens and suggests ecologically sustainable solutions.

Wu, Norbert. **Life in the Oceans.** (Planet Earth Series) New York: Little, Brown, 1991. 96 pp.

Part of the Planet Earth series, this oversized volume first shows the reader the many wonders of the sea through beautiful, full color pictures, emphasizing the importance of the ocean to life on earth. The final section of the book then explores the many angers that threaten the health of the oceans and its inhabitants, giving good general information about the various types of pollution.

Sewage

Bascom, Willard. **"The Disposal of Waste in the Ocean."** *Scientific American* 231, #14, (August 1974): pp. 16–25.

This article disputed the growing evidence that all ocean dumping is harmful, asserting: "If the process is thoughtfully controlled, it will do no damage to marine life . . ." This may have been one of the last published arguments of this nature.

Hoyle, Russ. **"The Dragon Lady."** *New York* 24 (April 8, 1991): pp. 74–79.

An in-depth look at the suit against E. Frank and General Marine Transport Corporation for illegal dumping in New York Harbor. One of the only sources of information on this key litigation in the area of marine dumping.

Ketchum, Bostwick H., Dana R. Kester, and P. Kilho Park., eds. **Ocean Dumping of Industrial Wastes.** New York: Plenum Press, 1981. 525 pp.

An important collection of essays that summarize the scientific information available at this critical time regarding the ocean's ability to assimilate waste dumped there intentionally. It issues a call for a full-scale research effort to better understand marine processes and "the capacity of these pelagic oceanic regions to assimilate wastes without detrimental effects." Technical, but seminal in nature.

Lynch, Kevin. **Wasting Away.** Michael Southworth, ed. San Francisco, CA: Sierra Club Books, 1991. 256 pp.

This book is a philosophical perspective on wastes. The role of waterways and the ocean in waste disposal throughout the centuries is explored in Chapter Two, *The Waste of Things*. The absurdity of the ways in which humans have chosen to deal with waste is eloquently stated in this sophisticated book about human nature and the planet Earth.

Marshall, Eliot. **"The Sludge Factor."** *Science* 242 (October 28, 1988): pp. 507–508.

Explains the Congressional ban on ocean dumping of sewage sludge that went into effect on January 1, 1992. One of the only lay interpretations of this ban available to date.

1991 Beach Pollution Report. Santa Monica, CA: Heal the Bay, 1991. 16 pp.

This report gives beaches in the Los Angeles area a "grade" based on the bacteria levels found in the water. It is intended to help determine swimming safety. It is a clear indictment of those who use the California coastline as a dumping ground.

Testing the Waters: A National Perspective on Beach Closings. New York: Natural Resources Defense Council, 1991.

Examines beach closings in 22 coastal states. Preparation overseen by Sarah Chasis.

Marine Debris

Blockstein, David E. **"Congress Tackles Ocean Plastic Pollution."** *Bioscience* 38 (January, 1988): p. 19.

An excellent description of the Senate ratification of Annex V of the International Convention for the Prevention of Pollution from Ships and P.L., 100-220, which bans the disposal of plastics in the sea within 200 miles of the U.S. coastline. Points out that the proliferation of plastics and the development of biodegradable substitutes has not yet been addressed.

Cleaning North America's Beaches: 1991 National Beach Cleanup Report. Washington, DC: Center for Marine Conservation, 1992. 110 pp.

Detailed report on findings of beach cleanups across the United States, Canada, and Mexico. Compiled based on the actual number of items in each category of debris found on the beaches. Reports for 1988, 1989, and 1990 are also available.

O'Hara, Kathryn, Suzanne Iudicello, and Rose Bierce. **A Citizen's Guide to Plastics in the Ocean: More Than a Litter Problem.** 3rd rev. ed. Washington, DC: Center for Marine Conservation, 1988. 143 pp.

A comprehensive view of the problems created by marine debris, with an overview of debris sources, impact on marine wildlife, a look at the legal front, a review of solutions, a guide to becoming involved in beach cleanup activities, and suggestions regarding prevention. Appendices include lists of federal agencies, Sea Grant offices, COASTWEEKS participants, international organizations, laws and treaties, and an excellent bibliography. One of the best books on the subject. Available through the Center for Marine Conservation.

O'Hara, Kathryn, N. Atkins, and Suzanne Iudicello. **Marine Wildlife Entanglement in North America.** Washington, DC: Center for Marine Conservation, 1986. 219 pp.

One of the first books published on the growing entanglement problem caused by marine debris, spearheaded by Katherine O'Hara, a leader in the field.

Persistent Marine Debris. University of Alaska: Alaska Sea Grant College Program, 1984. 41 pp.

Marks one of the first Sea Grant program efforts in the area of marine debris. Written in clear, understandable language, this brochure describes the extent and sources of marine debris, identifies how debris affects wildlife as well as humans, and recommends ways in which the federal government and others can work to alleviate the problem.

Weisskopf, Michael. **"Plastic Reaps a Grim Harvest in the Oceans of the World."** *Smithsonian* 18 (March, 1988): pp. 58–67.

A comprehensive look at the impact of plastics in the ocean, and at legislation designed to reduce the amount of plastic in the marine environment. An important addition to the literature because it includes arguments from both sides of the issue—manufacturers and environmentalists.

Toxic Chemicals

Dickson, David. **"Mystery Disease Strikes Europe's Seals."** *Science* 241 (August 19, 1988): pp. 893-895.

Explores the impact of toxic chemicals on seal immune systems and the viral infection that was affecting 90 percent of seal pups in the Wadden Sea off the coast of the Netherlands. A September 9th *Science* article (page 1284) entitled "Canine Distemper May be Killing North Sea Seals," is a good follow-up piece on this problem. Data on this tragedy is not yet available in book format.

Dold, Catherine. **"The New Guardians."** *American Health* 9 (September, 1990): pp. 44–48+.

Describes the battle waged by Tom Billecci (*See* Chapter Three) against his former employer Unocal from 1984-1990, and how the suit was finally won, resulting in a $4.22 million penalty against Unocal. Billecci was trying to force compliance with the Clean Water Act. Unocal had permit violations at its wastewater treatment plant at its San Pablo Bay refinery. Article includes information on water quality watchdogs, the hazards of water pollution, and how to take action on the environment.

Dunlap, Thomas R. **DDT: Scientists, Citizens, and Public Policy.** New Jersey: Princeton University Press, 1981. 318 pp.

This valuable volume traces pest control back into the 1800s, explaining the economic and bureaucratic reasons for the shift from the use of natural predators to the application of toxic chemicals. It documents both the warning and the fanfare regarding DDT and related pesticides. The volume is extremely well-documented with prolific footnotes and an extensive bibliography.

Heavy Metals

Furness, Robert W. and Philip S. Rainbow (eds.). **Heavy Metals in the Marine Environment.** Boca Raton, FL: CRC Press, 1990. 256 pp.

This book is highly technical in tone and content. It constitutes one of the best sources of information on natural leads and sources of heavy metals in the marine environment and how they interact with other substances. Bibliography and index.

Oil

Cahan, V. **"To Stop Spills, Punishment Must Cost More Than Prevention."** *Business Week* (July 10, 1989): 26.

A very practical, economic approach to preventing marine pollution by oil spills.

Davidson, Art. **In the Wake of the *Exxon Valdez:* the Devastating Impact of the Alaska Oil Spill.** San Francisco: Sierra Club Books, 1990. 333 pp.

A well-written, blow-by-blow account of the *Exxon Valdez* disaster. Traces the events leading up to and exacerbating the incident, as well as the web of disasters that followed the tanker accident.

Gill, C., F. Booker, and T. Soper. **The Wreck of the *Torrey Canyon*.** Newton Abbot (Devon): David and Charles, 1967. 128 pp.

One of the original accounts of this disaster which served as a wake up call to the citizens of European nations regarding the dangers inherent in petroleum transport.

Gist, Ginger L. **"The New Dead Sea: Persian Gulf Oil Spill Advances Ecological Clock."** *Journal of Environmental Health* (Spring, 1991): pp. 20–22.

One of the only updated sources of information about the impact of the ecological warfare that took place during the War in the Persian Gulf.

International Tanker Owners Pollution Federation (ITOPF). **Response to Marine Oil Spills.** London: Witherby & Co. Ltd., 1987. 150 pp.

A comprehensive review of the problems posed by marine oil spills and the response measures that can be implemented, from the viewpoint of the petroleum shipping industry. The five individual sections were first published in 1986 to supplement the information contained in the ITOPF oil spill training video series (Videotel Marine International) of the same name. Designed for those involved in training programs, contingency planning, and oil spill response. Available in English, French, and Spanish.

Keeble, John. **Out of the Channel: the *Exxon Valdez* Oil Spill in Prince William Sound.** New York: Harper Collins Publishers, 1991. 290 pp.

A scathing indictment of Exxon and the petroleum industry in general, footnoted and accompanied by facts and accurate maps. Keeble brings his expertise as a novelist to the task, without sacrificing data or documentation.

Nature Conservancy Council. **Oil Pollution Manual.** London: Nature Conservancy Council, 1977. 45 pp.

Guidelines for use by nature conservation and animal welfare organizations during an oil or other pollution incident.

Nelson-Smith, A. **Oil Pollution and Marine Ecology.** New York: Plenum Press, 1973. 260 pp.

Covers marine ecology in general, and then addresses specifically the nature of oil spills and their impact on wildlife.

Oil Spills: Just a Cost of Doing Business. Washington, DC: The Wilderness Society, May 1991. 11 pp.

This pamphlet lists all major oil spills in the United States from 1991 through 1992, and gives a brief description of each. It also quotes several Exxon executives, and provides useful statistics on the probability and size of oil spills.

Omohundro, John T. **Oil Spills: A Coastal Resident's Handbook.** Information Bulletin 164, Physical Sciences, Sea Grant 2. Ithaca, NY: New York State College of Agriculture and Life Sciences, a statutory college of the State University of New York at Cornell, [n.d.]. 14 pp.

This pamphlet defines an emergency oil spill and spells out appropriate citizen responses. It also lists federal and state government responsibilities and the relevant laws, and offers a list of emergency phone numbers.

Pearce, Fred. **"Wildlife Choked by World's Worst Oil Slick."** *New Scientist* 129 (February 2, 1991): pp. 24–25.

Although the focus is on wildlife, this article takes a look at the overall environmental impact of the oil slick on the Persian Gulf created during the War in Kuwait/Iraq. A map is included.

Spencer, Page. **White Silk & Black Tar: A Journal of the Alaska Oil Spill.** Minneapolis, MN: Bergamot Books, 1990. 150 pp.

A poignant, personal journal perspective on the *Exxon Valdez* oil spill in Prince William Sound, Alaska by a young scientist with the National Parks Service. Recounts two emotional months of 12- to 18-hour days on cleanup and assessment duty, providing a unique, human look at the most devastating oil spill so far in North America.

Steinhart, Carol E. and John S. **Blow Out: A Case Study of the Santa Barbara Oil Spill.** Belmont, CA: Duxbury Press, 1971. 135 pp.

A nontechnical, indignant account of offshore drilling gone awry—all over some of California's loveliest beaches, killing thousands of seabirds and permanently impacting the Santa Barbara ecosystem. Some very hard questions are asked, making this a thoughtful book about a tough subject.

Radioactive Materials

Cohen, Bernard L. **Before It's Too Late: A Scientist's Case FOR Nuclear Energy.** New York: Plenum Press, 1983. 292 pp.

This book is rather heavy reading, but it is one of a very limited number that promotes the use of nuclear power. The author forcefully defends the nuclear power industry's ability to safely dispose of high-level radioactive waste (Chapter 5), and discounts the dangers of low-level waste (Chapter 6). The author proposes that all nuclear waste be converted into glass and dropped "into the ocean at random locations." He claims this would "do no harm to ocean ecology."

Grossman, Karl. **Cover Up: What You Are Not Supposed To Know about Nuclear Power.** Sagaponack, NY: The Permanent Press, 1980. 312 pp.

Written by an investigative journalist, this compelling book maintains the confrontational tone of its title, reproducing original classified documents to support the author's vehement opposition to the nuclear power industry. Chapter Two explains how a nuclear power plant functions, Chapter Three details many accidents, several of which contaminated the marine environment, and Chapter Five investigates the full spectrum of radioactive waste disposal plans (prior to 1980), including a brief look at ocean dumping, as well as accidents that have occurred at waste disposal sites. Although marine pollution by radioactive substances is not specifically addressed, this book is a must read for those who suspect they've been denied complete information about radioactive waste disposal.

Shapiro, Fred C. **Radwaste.** New York: Random House, 1981. 288 pp.

A comprehensive look at radioactive wastes of every kind. Includes a great deal of valuable history. The story of ocean dumping from 1946 through the 1970s is told on pages 122–127, including significant covert dumping. Uses nontechnical language in a reader-friendly format.

Weiss, Ann E.. **The Nuclear Question.** New York: Harcourt Brace Jovanovich, 1981. 160 pp.

In addition to Chapter Five, entitled *The Broken Cycle,* which explains in layman's terms the back end of the nuclear fuel cycle and the controversy surrounding both reprocessing and disposal of high and low level radioactive waste, two other chapters discuss various nuclear power plant accidents and their potential environmental impact. This book provides the level of understanding needed to come to personal conclusions regarding the safety of nuclear power plants and their impact on the marine environment.

7

Selected Nonprint Resources

Audio and Visual

This section includes audiocassettes and tapes, videotapes in VHS, Beta, and 3/4 inch U-matic formats, 16 mm films, film strips, and slide shows, listed in alphabetical order by title. All are available for rent or purchase, and a source for each is listed.

Alaska: Crude Awakening
Type: VHS or Beta, color
Length: 47 minutes
Date: 1990
Cost: Purchase $179; rental $75
Source: Films for the Humanities & Sciences
P.O. Box 2053
Princeton, NJ 08543-2053
(800) 257-5126 or 609-452-1128
Fax: 609-452-1602

This CBS *48 Hours* program goes beyond the *Exxon Valdez* oil spill to explore the impact of oil and mineral exploitation on the economy of Alaska and the lifestyles of Alaskans.

America's Biggest Oil Spill
Type: VHS video
Length: 50 minutes

Date: Copyright 1989, released 1990
Cost: $19.95
Source: Summit Media Company
27811 Hopkins Ave., Unit 1
Valencia, CA 91355
(800) 777-8668

This TV journalism treatment of the *Exxon Valdez* oil spill terms the accident an American Tragedy and compares the ecological damage to that inflicted on Hiroshima by the atomic bomb during World War II. This bleak outlook, visually demonstrated by shots of oiled birds and beaches, is followed up by a 1990 update segment. The economic windfall for local citizens—and for Exxon itself—is highlighted. The court settlement with Exxon occurred after the release of the film and is not covered. This film should generate lively discussion about the economics of ecology.

America's Trash Hits Home
Type: VHS, Beta and 3/4 U-matic, color
Length: 30 minutes
Date: 1988
Cost: $160
Source: ABC Distribution Co.
Capitol Cities/ABC Video Enterprises
825 7th Avenue
New York, NY 10019
(212) 887-1725

Ted Koppel takes a look at the problem of waste in America on the television program *Nightline*. Particular emphasis is placed on the hazardous wastes washing up on Atlantic coast beaches.

Are You Swimming in a Sewer?
Type: VHS and 3/4-inch U-matic with teacher's guide; color & some B&W
Length: 58 minutes
Date: 1986 (VHS–1987)
Cost: $99 (purchase only)
Source: WGBH-Nova
125 Western Ave.
Boston, MA 02134
(617) 492-2777

This excellent *Nova* segment reveals that 5.5 billion gallons of wastewater flow into the ocean every day from American city sewer systems and industrial discharge pipes. It explains why scientists are reconsidering assumptions about the ocean's ability to dilute waste and encourages viewers to decide whether the cost of switching to nonpolluting methods of waste disposal is a worthy investment in the future of their coastal waters. Closed-captioned.

Baymen, Our Waters Are Dying
Type: 16 mm (video—inquire)
Length: 29 minutes
Date: 1977
Cost: Purchase $535; rental $50
Source: Sea Horse Films
12 Harrison Street
New York, NY 10013
(212) 226-0294

This award-winning documentary paints a vivid portrait of the clam-diggers of East Long Island, who harvest shellfish from the area's sea-coast bays. It shows how and why marine pollution threatens their traditional way of life.

Between the Devil and the Deep Blue Sea
Type: VHS video
Length: 32 minutes
Date: 1990
Cost: $250
Source: Landmark Films
3450 Slade Run Drive
Falls Church, VA 22042
(800) 342-4336

The oil spill of the *Exxon Valdez* and its consequences on Alaska's Prince William Sound is the subject of this environmentally aware film.

The Big Spill
Type: 1/2 in., VHS, color
Length: 58 minutes
Date: 1990
Cost: $99 (purchase only)
Source: Coronet Film & Video
108 Wilmot Rd.
Deerfield, IL 60015
(800) 777-8100

An excellent NOVA presentation on the *Exxon Valdez* oil spill in Alaska, one that portrays our limited ability to prevent, contain, or mitigate such massive oil spills. Computer simulations, dramatic reenactments of the accident, and on-site footage combine to create a powerful piece of scientific journalism. Conflicting points of view on every aspect of the spill, from tanker design to beach cleanup methodologies, are given equal time and treatment. This creates a balanced picture of the spill which leaves the viewer with enough information to develop a personal opinion regarding the event.

Black Water Time Bomb

Type: VHS, Beta, 3/4 U-matic
Length: 30 minutes
Date: 1984
Cost: Purchase, rent/lease
Source: San Diego State University
Learning Resource Center
San Diego, CA 92182
(619) 594-5200

A vivid documentary on the environmental effects of the millions of tons of untreated sewage dumped in the Pacific Ocean by the city of Tijuana, Mexico. A good case study on the dynamics of ocean sewage disposal as it takes place in coastal areas around the world.

Chesapeake: Living Off the Land

Type: VHS & 16 mm
Length: 28 minutes
Date: 1992
Cost: Free loan; purchase $29 (or $15 for educational organizations)
Source: Save the Bay Shop
162 Prince George Street
Annapolis, MD 21401
(410) 268-8832.

An excellent overview of the past, present, and future of Chesapeake Bay. The film presents not only the many environmental problems that beset this extremely important estuary, but also some possible solutions to those problems. Along with the exquisite footage are recommendations on how everyone can help preserve the largest estuary in the United States. Recipient of a CINE Award; hosted by Walter Cronkite. Use in conjunction with **Turning the Tide,** by Tom Horton (also available through the Save the Bay Shop).

Coastal Clean-Up

Type: Slide show (includes a script)
Length: 55 slides
Date: 1990
Cost: Purchase $25; free 5-week loan

Source: Center for Marine Conservation
NOAA Marine Debris Information Office
1725 DeSales, NW, Suite 500
Washington, DC 20036
(202) 429-5609

How to conduct a beach cleanup and collect data. This slide show is designed to help school and community groups organize beach cleanups.

Cousteau Collection
Type: VHS, Beta
Length: Approx. 48 minutes each
Date: Varies
Cost: $19.99
Source: Turner Home Entertainment Company
P.O. Box 105366
Atlanta, GA 30348
(404) 827-3066

In volume one of this six-volume series (**Alaska: Outrage at Valdez**), Jean-Michel Cousteau and his crew study the serious impact of the *Exxon Valdez* catastrophe on Alaska's fragile ecology. In Volume two, **Lilliput in Antarctica,** Jacques Cousteau takes six children (representing six continents) to Antarctica to demonstrate to the world that this frozen land must be preserved for future generations. The children see the beauty of the land and the ugliness of the oil still seeping from the *Bahia Paraiso,* a tanker that ran aground in 1989. Volume four, **Tahiti: Fire Waters,** documents the long-term impact on the inhabitants of the Tahitian islands of radioactivity from nuclear experiments. The other volumes of this series do not deal with marine pollution.

Cousteau Odyssey
Type: VHS, Beta
Length: About 58 minutes each
Date: Varies
Cost: $24.98
Source: Warner Home Video
4000 Warner Blvd.
Burbank, CA 91522
(818) 954-6000

Volumes five and seven of this ten-volume series deal directly with marine pollution. In volume five (**Time Bomb at 50 Fathoms**—1978), Jacques Cousteau and his crew aboard the *Calypso* discover 900 sunken drums of poisonous chemicals off the coast of Italy. The salvaging operation is documented as a race that must be completed before the drums begin to leak. In volume seven (**Mediterranean: Cradle or Coffin**—1980), Cousteau and his crew spend half a year examining the effects of industry and modernization on the Mediterranean Sea, which they find to be in grave ecological jeopardy.

The Cry of the Beluga

Type: 3/4 U-matic cassette video
Length: 51 minutes
Date: 1989
Cost: Purchase $390; rent/lease also available
Source: Icarus Films
153 Waverly Place, 6th Floor
New York, NY 10014
(212) 727-1711

A well-rounded look at how our modern technological society has affected the beluga whale population found in the St. Lawrence Seaway. Pollution from substances such as PCBs has led to cancer rates fifteen times higher than normal in the beautiful belugas of the St. Lawrence. Efforts are being made by conservationists to protect the whales.

Danger at the Beach

Type: VHS, color
Length: 60 minutes
Date: 1991
Cost: Purchase $49.95; rental $8.50
Source: PBS Video
1320 Braddock Place
Alexandria, VA 22314
(800) 424-7963, or (800) 328-7271

A vivid depiction of the growing danger of toxic waste contamination of U.S. coastal waters. This video shows that, although nearly 75 percent of Americans live within 50 miles of the coastline, many still choose to ignore the threat posed by the pollution of the coastal waters. A National Audubon Society film narrated by Ted Danson.

Don't Mess With Texas Beaches

Type: VHS, color
Length: 23 minutes
Date: N/D
Cost: $20
Source: Sea Grant College Program
Texas A&M University at Galveston
P.O. Box 1675
Galveston, TX 77553-1675
(409) 762-9800

Focuses on the international costs of ocean dumping and its harmful effects on marine life as well as on the Texas economy. Information on the cost of beach cleanups is also included.

Don't Teach Your Trash to Swim
Type: Cassette tape and video tape
Length: 30-second audio clip and 2–3-minute music video
Date: 1991
Cost: Free (please provide a blank audio and video tape)
Source: Reef Relief
P.O. Box 430
Key West, FL 33041
(305) 294-3100

This rap song was developed for use as a public service announcement to discourage people from littering beaches and waterways.

Do We Really Want To Live This Way?
Type: VHS
Length: 60 minutes
Date: 1990
Cost: $29.95 (purchase only)
Source: The Annenberg CPB Project
P.O. Box 2345
S. Burlington, VT 05407
(800) LEARNER

Segment two of the *Race to Save the Planet* series uses Los Angeles and the Rhine river as two dramatic examples of the environmental cost of progress. Ways to sustain the environment in Western industrialized areas are discussed.

Down the Shore
Type: VHS, 3/4 U-matic
Length: 60 minutes
Date: 1989
Cost: $59.95 (purchase only)
Source: PBS Video
11858 La Grange Ave.
Los Angeles, CA 90025
(213) 820-0991

This well-made video takes a close look at both legal and illegal dumping of toxic wastes off the shores of the United States. The scope of the problem and its repercussions are examined.

The Drowning Bay
Type: VHS, Beta, 3/4 U-matic

Length: 9 minutes
Date: 1970
Cost: Purchase (VHS) $135.00; rental $33 ($16.50 each day thereafter)
Source: Phoenix/BFA Films
468 Park Ave., S.
New York, NY 10016
(800) 221-1274

Although this documentary is more than 20 years old, its message remains important. It vividly illustrates the ecological problems experienced by a city which has allowed pollution to destroy a major part of a famous natural resource. It is short and to the point.

500 Million Years Beneath the Sea
Type: Video and 16 mm
Length: 48 minutes
Date: 1973
Cost: Purchase video—$99; 16 mm—$16.50, plus $5.00 shipping; free rental.
Source: Purchase:Churchill Films
662 N. Robertson Blvd.
Los Angeles, CA 90069-9990
(213) 657-5110
Rental: Florida State University at Tallahassee
Instructional Support Center Film Library
54 Johnston Building
Tallahassee, FL 32306-1019
(904) 644-2820

Reveals the ecological imbalance caused by pollution in a South Pacific lagoon. Some species have been destroyed and an adaptive sea snake developed in this body of water. This is now an older documentary, but it was ahead of its time when it was originally released during the early 1970s.

From Sea to Shining Sea
Type: VHS, Beta, 3/4 U-matic
Length: 20 minutes
Date: 1986
Cost: $95
Source: Bullfrog Films, Inc.
P.O. Box 149
Oley, PA 19547
(800) 543-3764

A documentary featuring the efforts of Greenpeace to halt the dumping of chemical toxins into the Atlantic Ocean in one small coastal town. This is the sort of battle Greenpeace has fought repeatedly, and this case study reveals the power of economic considerations and progress over logic and the rights of future generations. A compelling story.

Living with the Spill

Type: VHS, color
Length: 52 minutes
Date: 1990
Cost: $390
Source: First Run/Icarus Films
153 Waverly Place, 6th Floor
New York, NY 10014
(212) 727-1711

A hard-hitting sociological study of the *Exxon Valdez* oil spill. This British production poignantly illustrates the ecological disaster by contrasting it with the natural beauty of Alaska, and also describes the transformation of the Alaskan community that took place when the "second spill" occurred: the $6 million spent during the first four months of cleanup. This is an emotional, rather than an objective, treatment of America's largest spill, one that is sure to generate discussion.

Marine Debris and Entanglement

Type: Slide show, color
Length: 47 slides
Date: 1986
Cost: Purchase $25; free 5-week loan
Source: Center for Marine Conservation
NOAA Marine Debris Information Office
1725 DeSales N.W., Suite 500
Washington, DC 20036
(202) 429-5609

This slide show provides an overview of the problems of wildlife ingesting or becoming entangled in marine debris. Produced by the Center for Marine Conservation, it is designed to educate upper elementary school-age to adult audiences regarding the hazards of plastic debris.

The Ocean Planet: The Death of the Mississippi

Type: VHS or Beta, color
Length: 23 minutes
Date: 1990
Cost: Purchase $149, rental $75

Source: Films for the Humanities & Sciences
P.O. Box 2053
Princeton, NJ 08543-2053
(800) 257-5126 or 609-452-1128
Fax: 609-452-1602

Details the decline of the great Mississippi River, from a wild and scenic waterway to an oily morass containing deadly toxins that wash freely and in great quantity into the Gulf of Mexico. Explains that it is not possible to pollute our rivers without fouling our seas. Part of *The Blue Revolution* series.

Ocean Pollution: What You Should Know
Type: Slide show
Length: 45 minutes
Date: N.d.
Cost: Free on loan
Source: Clean Ocean Action
P.O. Box 505
Sandy Hook
Highlands, NJ 07732
(908) 872-0111

Presents a good overview of the various types of ocean pollution, especially coastal marine debris, and describes a number of solutions.

The Ocean Sink
Type: VHS or Beta, color, 1/2 inch
Length: 29 minutes
Date: 1990
Cost: Purchase $149, rental $75
Source: Films for the Humanities & Sciences
P.O. Box 2053
Princeton, NJ 08543-2053
(800) 257-5126 or 609-452-1128
Fax: 609-452-1602

Examines various places around the world where the sea has been used as a dumping ground for highly toxic substances, including Minamata, Japan. The difficulties in disposing of these substances, even when the problem is fully recognized, are also explored in detail. Part of *The Blue Revolution* series.

Passion & Commitment
Type: Audio cassette
Length: 15 minutes
Date: 1987

Cost: $1.00
Source: Windstar Foundation
2317 Snowmass Creek Road
Snowmass, CO 81654
(303) 927-4777

The address given by Jean-Michel Cousteau at the second annual Evening of Choices for the Future Symposium to address the quality of life on earth. An impassioned plea to save the oceans in order to save the planet.

Protecting Coastal Waters: A Community Approach
Type: VHS
Length: 18 minutes
Date: 1992
Cost: Loan only: $10 per week
Source: New England Interstate Environmental Training Center (NEITC)
2 Fort Road
South Portland, ME 04106
(207) 767-2539

This documentary, produced to inform students, coastal communities, and resource managers about coastal pollution issues, focuses on sewage outfalls, sludge, and marine debris.

Remnants of Eden
Type: VHS
Length: 60 minutes
Date: 1990
Cost: Purchase $29.95
Source: The Annenberg CPB Project
P.O. Box 2345
S. Burlington, VT 05407
(800) LEARNER

Segment five of the excellent *Race to Save the Planet* series. The Florida Everglades is the marine example (there are also four nonmarine examples) of how the diversity of living organisms can be protected while addressing the needs of growing human populations. Produced by WGBH TV (Boston).

The Sea: Mysteries of the Deep
Type: VHS
Length: 22 minutes
Date: 1979
Cost: Purchase $79; free rental

Source: Purchase: Britannica Films
310 S. Michigan Ave.
Chicago, IL 60604
(800) 554-9862
Rental: Pennsylvania State University
University Park, PA 16802
(814) 863-4396

Photographed from a special observation bubble on the SSP *Kaimalino,* this video examines some of the inhabitants of the ocean depths, focusing on the predator-prey relationships that abound undersea, particularly on the recent but deadly predators that threaten all sea life—human beings.

Seas and Oceans—An Evening of Choices for the Future Symposium
Type: Audio cassette, VHS video
Length: Audio: 15 minutes; video: 49 minutes
Date: 1986
Cost: Audio: $1.00; video: $9.95
Source: Windstar Foundation
2317 Snowmass Creek Road
Snowmass, CO 81654
(303) 927-4777

The address given by Jean-Michel Cousteau at the first annual Evening of Choices for the Future Symposium, when receiving the Windstar Award on behalf of his father. Discusses the dangers marine pollution presents to the health of the planet. An inspiring speech.

Seas under Siege
Type: VHS or Beta, color, 1/2 inch
Length: 56 minutes
Date: 1988
Cost: Purchase $179, rental $75
Source: Films for the Humanities & Sciences
P.O. Box 2053
Princeton, NJ 08543-2053
(800) 257-5126 or 609-452-1128
Fax: 609-452-1602

Takes a look at the sources of marine pollution, including industrial effluents; by-products and waste; storm runoff from land treated by herbicides and from oil slick-covered highways; and sewage and toxic wastes. Also explores some of the consequences of this pollution, using fouled New Jersey beaches and dead Baltic seals to deliver a strong message. The program ends with a dire warning: "This time it was seals; next time it could be the people living along the coast." A powerful piece.

Sentinel of the Sea

Type: Video
Length: 16 minutes
Date: 1988
Cost: Available on loan
Source: Clean Seas
1180 Eugenia Place, Suite 204
Carpinteria, CA 93103
(805) 684-3838

A film for use by oil industry personnel, government agency personnel, and wildlife rehabilitation personnel in training individuals how to undertake beach cleanups, handle oiled birds, and generally protect the seas and its inhabitants.

Texas Shores—Saving What's Left

Type: VHS, color
Length: 26:39
Date: N.d.
Cost: $20
Source: Sea Grant College Program
Texas A&M University at Galveston
P.O. Box 1675
Galveston, TX 77553-1675
(409) 762-9800

The costs of ocean dumping are clearly laid out, both in terms of the loss of valuable marine life and in the expense of regular beach cleanups that are needed to make coastal areas safe for beach goers. A good review of marine debris, presented as an economically unsound practice.

Trashing the Oceans & Port of Newport (Oregon)

Type: Video
Length: 14 minutes
Date: 1988
Cost: Purchase: $10; free one-month loan
Source: Center for Marine Conservation
Marine Debris Information Office
1725 DeSales N.W., Suite 500
Washington, DC 20036
(202) 429-5609

Describes the problems of marine debris and lists a number of solutions. The fishing port of Newport, Oregon is then used as a case study to demonstrate the problems caused by marine debris and to show how they can be solved.

Treasures of Neptune: Klondike on the Ocean Floor.
Type: VHS or Beta, color
Length: 26 minutes
Date: 1990
Cost: Purchase $149; rental $75
Source: Films for the Humanities & Sciences
P.O. Box 2053
Princeton, NJ 08543-2053
(800) 257-5126 or 609-452-1128
Fax: 609-452-1602

Looks at the impact on marine ecosystems of a number of new technologies (from industrial drilling rigs to giant vacuum cleaners) used to recover the valuable resources lying on and under the ocean floor. The relationship between plate tectonics and deep sea mineral deposits is also explained, and the ways in which the ocean floor is being mapped are explored. Part of *The Blue Revolution* series.

Troubled Waters
Type: VHS, 3/4 U-matic
Length: 28 minutes
Date: 1984
Cost: $340
Source: University of California at Berkeley Extension Media Center
2176 Shattuck Ave.
Berkeley, CA 94704
(510) 642-0460

A well-done documentary that examines the many ways in which oil exploration is threatening California's coast and coastal wildlife.

Turning the Tide: Keeping Pollution at Bay
Type: VHS
Length: 30 minutes
Date: 1992
Cost: Purchase $25; loan $10 per week.
Source: New England Interstate Environmental Training Center (NEITC)
2 Fort Road
South Portland, ME 04106
(207) 767-2539

Explores the pollution problem in Buzzard's Bay, Massachusetts. Describes the efforts of individuals and the local government in restoring and protecting the bay. Could be used as a case study and applied to other bodies of water.

Underwater Marine Debris Collection

Type: Slide show
Length: 45 slides (includes script; free color brochure also available)
Date: 1992
Cost: Purchase: $35; loan: one week free
Source: Center for Marine Conservation
NOAA Marine Debris Information Office
1725 DeSales N.W., Suite 500
Washington, DC 20036
(202) 429-5609 or (813) 895-2188

Developed to teach certified scuba divers how to clean up underwater marine debris (especially monofilament fishing line) from living organisms such as coral and sponges. It is important to use the techniques shown to remove underwater debris to prevent undue damage to marine organisms. This slide show also teaches divers what to leave behind and what to clean up while underwater. A free, four-panel, full-color brochure is also available.

Waste Not, Want Not

Type: VHS
Length: 60 minutes
Date: 1990
Cost: Purchase: $29.95
Source: The Annenberg CPB Project
P.O. Box 2345
S. Burlington, VT 05407
(800) LEARNER

Demonstrates ways to generate less waste, using concepts developed around the world. The video also explains that the less waste that is produced, the less likely it is that waste will inadvertently end up in the marine environment. Recycling and waste treatment are discussed. This constitutes Segment eight of the excellent *Race To Save the Planet* series.

What Price Progress?

Type: 16 mm film
Length: 48 minutes
Date: 1977
Cost: Rental $23 per week

Source: University of Illinois Film Center
1325 S. Oak St.
Champaign, IL 61820
(800) 367-3456

Three cases of industrial pollution are documented in this Jacques Cousteau television program: the poisoning of Minamata Bay (Japan) is described in careful detail, explaining the nature of methyl mercury poisoning and the nearby factory's unwillingness to take responsibility for the tragedy. Industrial pollution in a Canadian river and a Minnesota bay are also documented.

Where Have All the Dolphins Gone?

Type: VHS, Beta, 3/4 U-matic
Length: 58 minutes
Date: 1990
Cost: $59.95
Source: The Video Project
5332 College Ave., Ste. 101
Oakland, CA 94618
(510) 655-9050

A spectacular documentary on the fate suffered by one type of marine mammal because of human activities. The fact that little is being done to save the dolphins is also made abundantly clear. This is a critically acclaimed film that is sure to spark an interest in preventing all types of marine pollution.

Whose Sea Is This?

Type: VHS or Beta, color
Length: 26 minutes
Date: 1990
Cost: Purchase $149, rental $75
Source: Films for the Humanities & Sciences
P.O. Box 2053
Princeton, NJ 08543-2053
(800) 257-5126 or 609-452-1128
Fax: 609-452-1602

Examines the political and economic problems surrounding "ownership" of the ocean and the role of the Law of the Sea in resolving these problems. Provides a good, although fairly superficial, explanation of the Law of the Sea, which establishes 12-mile exclusive economic zones (EEZs) for all coastal nations, and tries to determine responsibility for the health of the oceans as a whole. Part of *The Blue Revolution* series.

Computer Networks and Databases

COASTNET
Oceans & Coastal Protection Division
Office of Wetlands, Oceans & Watersheds
Office of Water (WH-556F)
U.S. Environmental Protection Agency
401 M Street, SW
Washington, D.C. 20460
(202) 260-8482 (access via modem only)

This electronic bulletin board system allows those with a computer, a modem, and proper software to exchange information and documents on coastal matters. COASTNET can also be accessed through EcoNet.

CompuServe

To obtain a CompuServe Membership Kit (contains user ID number and a temporary password) check local computer retailers or call 1-800-848-8199. Started in 1969, CompuServe serves 500,000 members worldwide. Environmental resources include: The Network Earth Forum, The Good Earth Forum, and others. Message Boards can be used to participate in or initiate discussions about marine pollution.

Dialog Information Services
Palo Alto, CA
(800) 334-2564

Many libraries subscribe to this online database service. In addition to the various subject-specific databases (i.e., Enviroline, Environmental Bibliography, Oceanic Abstracts, Pollution Abstracts) provided through Dialog that are relevant to marine pollution, there are also a number of newspaper databases, where subject searches can be used to retrieve newspaper articles on events related to marine pollution.

Directory of Environmental Organizations
Coverage: International
File Size: 3,000 entries
Updates: Annual
Provider: Ecology Center of Southern California
Educational Communications, Inc.
P.O. Box 351419
Los Angeles, CA 90035
Format: Labels or diskette: $200 (also available in print: $30)

This database includes information on more than 3,000 organizations concerned with the environment. It has a definite advocacy slant and focuses primarily on organizations in the United States (many in California). Only about 10 percent work specifically in the field of marine pollution, but many undertake periodic projects that have to do with the degradation of the oceans. Contains a great deal of information not available elsewhere.

EcoNet
Institute for Global Communications
18 De Boom St.
San Francisco, CA 94107
(415) 442-0220

A sophisticated, nonprofit global telecommunications system supported by corporations and foundations. A fee of just $15 provides individuals with an EcoNet identification number and a user's manual. The system enables those who have a computer, a modem, and telecommunications software to send messages, letters, or documents to any participating organization or individual for just pennies a page. Teleconferencing and database sharing is also possible. This is the most extensive environmental computer network in the world. Currently has more than 1,800 users, including environmental organizations with a collective membership of more than 11 million people.

Enviroline
Coverage: 1971–present
File Size: 144,225 records
Updates: Monthly
Provider: R. R. Bowker
New York, NY

Covers the world's environmental information via 5,000 primary and secondary source publications in fields such as management, law, and biology, as they relate to environmental issues. Available at libraries through Dialog.

Environmental Bibliography
Coverage: 1973–present
File Size: 398,698 records
Updates: Bimonthly
Provider: Environmental Studies Institute
Santa Barbara, CA

Cites references to articles in 300 periodicals covering the fields of general human ecology, water resources, etc. Not specific to marine pollution, but can be searched using key words and phrases such as marine pollution, ocean studies, oil pollution, and marine debris. Available at libraries through Dialog.

Oceanic Abstracts

Coverage: 1964–present
File Size: 208,869 records
Updates: Bimonthly
Provider: Cambridge Scientific Abstracts
Bethesda, MD

Organizes and indexes technical literature published worldwide on marine subjects in journals, books, technical reports, conference proceedings, and government and trade publications. Major subject areas covered include marine biology, marine pollution, and ships and shipping. Available at libraries through Dialog.

Oil Spill Intelligence Report

Publisher: Cutter Information Group
Frequency: Weekly
Available through NewsNet (800-345-1301) Code-EV32

The only weekly source of global information on oil spill cleanup, control and prevention.

Pollution Abstracts

Coverage: 1970–present
File Size: 157,553 records
Updates: Bimonthly
Provider: Cambridge Scientific Abstracts
Bethesda, MD

A leading reference source for highly technical, scientific literature on environmental pollution, its sources, and its control. Covers environmental quality, marine pollution, radiation, solid wastes, pesticides, etc. Available at libraries through Dialog.

SEAPRESS
Seas At Risk

Vossiusstraat 20-lll
NL-1071 AD
Amsterdam
The Netherlands
Phone: 31-20-675-4336
Fax: 31-20-675-3806

An information bureau that publishes press releases on marine environment developments and provides the media with background information. Focused on Europe, but coverage is international in scope.

The WELL
Whole Earth Catalog/Review
San Fransisco, CA
(415) 332-4335 (phone)
(415) 332-6106 (modem)

The "Whole Earth 'Lectronic Link" is a "conferencing" network composed of a series of discussions on a wide range of issues, including marine pollution. It specializes in "online conversations." The Environment Conference includes options of offshore drilling, "Save Our Seas-South Pacific," etc.

Television and Radio Series

EcoNews
Type: VHS, 1/2 inch and 3/4 inch videocassette
Length: 30 minutes
Date: Currently being aired
Cost: Purchase \$30–70; rental \$15–20, plus deposit
Source: Educational Communications, Inc.
P.O. Box 351419
Los Angeles, CA 90035
(310) 559-9160

Weekly television series available via public access, cable, and PBS stations, as well as on videocassette for rent or purchase. Topics cover the ecology spectrum, including coastal marine life, oil drilling in marine waters, pollution in Santa Monica Bay, plastic recycling, the *Exxon Valdez* oil spill, ocean trash and global warming, and pollution in the Puget Sound. Index of topics available.

Environmental Directions
Type: Radio programs (available on audio cassettes)
Length: 30 minutes
Date: Currently being aired
Cost: \$15
Source: Educational Communications, Inc.
P.O. Box 35473
Los Angeles, CA 90035
(310) 559-9160

More than 850 shows cover a variety of ecological issues, although only about 20 percent deal with marine pollution specifically. Primarily interview format, covering subjects of either broad or highly topical interest. Index of topics available.

E-Town

Type:	Weekly radio program
Length:	56 minutes
Date:	Currently being aired
Cost:	Not available on cassette or as transcript; broadcast on National Public Radio member stations
Source:	E-Town Productions P.O. Box 954 Boulder, CO 80306 (303) 443-8696

A musical variety show with an environmental message, produced live in Boulder, Colorado, and aired on public radio stations across the nation. Frequently covers marine pollution topics. Accepts nominations for the E-chievement awards. Recently received an Environmental Protection Agency grant designed to support unusual approaches to environmental education.

Radio Earth Island

Type:	Radio program
Length:	Varies
Date:	Currently being aired
Cost:	N/A
Source:	Earth Island Institute 300 Broadway, Suite 28 San Francisco, CA 94133 Phone:(415) 788-3666 Fax:(415) 788-7324

This innovative and topical environmental radio program series is produced and distributed via satellite to radio stations internationally. Not specifically focused on marine pollution issues, but about 20 percent of the programs are relevant to this topic.

Glossary

accumulation: The process whereby quantities of toxins present in the bodies of marine organisms increase very quickly as the food chain progresses.

absorption: The process by which a dissolved substance enters into the inner structure of another substance or organism. There are two ways this can occur: physiologically, as when nutrients enter the bloodstream through the intestinal walls; and physicochemically, as when water takes up molecules of gas.

adsorption: The process by which a gaseous, dissolved, or particulate substance attaches itself to the surface of another substance, as when oil attaches itself to particles of solid matter.

ALARA: As low as reasonably achievable. A pragmatic approach to determining the 'allowable' amounts of pollutants in the environment.

anaerobic decomposition: Decomposition without the presence of oxygen, by bacteria that require no oxygen to survive (known as anaerobic bacteria).

anthropogenic: Human-made, or resulting from the activities of humans.

assimilative capacity: The presumed ability of the ocean to harmlessly digest wastes and toxins by reducing their strength (*See* **dilution**), spreading them through the biosphere (*See* **dispersion**), by the settling action that sends them to the ocean floor (*See* **sedimentation**), and by breaking them down chemically (*See* **degradation**).

b/d: Barrels per day.

background radiation: The natural level of radiation, excluding that generated by human activity.

becquerel (Bq): A measure of radioactivity. For a given amount of radioactive substance, it is the number of atomic disintegrations per second.

bioaccumulation: The retention and cumulative increase of heavy metals, **halogenated hydrocarbons,** and other pollutants in plant and animal tissues.

biodegradation: The process in which substances are broken down into simpler compounds by living organisms, especially bacteria and microfungi. At least 90 strains of microorganisms are capable of biodegrading some components of petroleum.

biogenic: Pertaining to or produced by plants or animals.

biomass: The total amount of living matter in an area or habitat.

biological oxygen demand (BOD): The amount of oxygen required to decompose organic matter. Used as a measure of the oxygen requirement of bacteria in a water sample, it is an index of the level of pollution by organic matter (typically sewage).

biomagnification: A result of bioaccumulation in the food chain. Animals that feed on bioaccumulators end up with a higher relative concentration of the toxins that are lodged in the tissues of the animals they feed on. The concentration and thus the impact of toxic substances is magnified as they are passed up the food chain.

bioremediation: The attempt to speed up the natural decomposition of oil by bacteria, through the introduction of nitrogen-based fertilizers. This method is a favorite of oil companies but is viewed skeptically by environmentalists. It was used experimentally during the *Exxon Valdez* cleanup.

biosphere: The portion of the earth's crust, waters, and atmosphere that supports life.

biosynthesis: The formation of chemical compounds by a living organism, either by synthesis or via degradation.

bioturbation: Disturbance of sediments by creatures living in or on the ocean bed.

bloom: An extreme overgrowth of phytoplankton. Can suffocate marine life and produce toxins that contaminate waters and shellfish.

blowout: The uncontrolled release of oil from an offshore well.

BOD: *See* **biological oxygen demand.**

burden: The total amount of a substance in all or part of the biosphere.

CFC: *See* **chlorofluorocarbon.**

chemical oxygen demand (COD): The amount of oxygen required to degrade nonliving substances in the sea through oxidation.

chlorinated hydrocarbons: Hydrocarbon molecules with chlorine atoms attached at specific locations. Chlorinated hydrocarbons are found in many pesticide residues. They retard growth and reproduction in marine organisms, and cause cancer. They are stored in fatty tissue and thus impact the food chain via **bioaccumulation** and **biomagnification.** *See also* **halogenated hydrocarbons.**

chlorofluorocarbon (CFC): Class of highly stable hydrocarbon molecules that have either chlorine or fluorine attached to them. CFCs are used mainly as refrigerants and cleaning solvents. When they enter the stratosphere, CFCs break down under strong ultraviolet radiation and react with ozone, converting it to oxygen. This diminishes the ability of the earth's ozone layer to protect living things from ultraviolet radiation.

chocolate mousse: *See* mousse.

COD: *See* **chemical oxygen demand.**

contamination: The presence of elevated concentrations of a toxic substance.

continental margin: The portion of the seabed adjoining the land, including the continental shelf and the continental rise.

CRISTAL: The Contract Regarding an Interim Supplement to Tanker Liability for Oil Pollution, a voluntary agreement among oil companies to provide supplemental compensation for oil pollution damage, beyond that which is legally obtainable elsewhere.

critical pathways: Routes by which radioactivity can spread throughout the biosphere. Used by scientists to gauge the risk of radiation to human beings.

CZMA: Coastal Zone Management Act, rewritten and passed by the U.S. Congress in 1990.

DDE: One of the most harmful by-products of **DDT.** This is a persistent toxin that bioaccumulates up the food chain.

DDT: Dichlorodiphenyltrichloroethane, a halogenated hydrocarbon. DDT is a potent insecticide that enters the marine environment from agricultural runoff into rivers, as well as through precipitation.

degradation: The breakdown of a compound. Used frequently to describe what happens to organic hydrocarbons in the marine environment.

depuration: The process by which bivalves such as clams and oysters cleanse themselves of toxic substances accumulated in contaminated waters when they are subsequently placed for a period of time in clean waters.

detoxification: The process by which the body changes toxins into less poisonous compounds or into substances that can be more easily excreted.

diffusion: In a liquid such as seawater, the movement of dissolved substances from areas of high concentration to areas of lower concentration.

dilution: The reduction in concentration of a substance by mixing with far larger amounts of another substance, usually water. The ocean was long considered to have a nearly boundless capacity for dilution of large quantities of wastes.

dioxin: A halogenated hydrocarbon consisting of benzene, chlorine, and oxygen. The deadliest type of dioxin is 2,3,7,8-tetrachlorodibenzo-*p*-dioxin, or TCDD. Six millionths of a gram of TCDD will kill a rat. Nonlethal doses cause cancer, liver damage, hair loss, birth defects, and damage to the immune system. Dioxins are a by-product of the process used to produce chemicals called chlorophenols and the products made from them, including wood preservatives and disinfectants; they are produced by pulp and paper mills that use a chlorine bleaching process. When paper products are put in a landfill or burned, dioxins are emitted into the air and carried to the sea. Dioxins can be broken down by sunlight, but they persist for years in soils and sediments, as well as in plants and body tissues.

dispersants: Detergent-like chemicals that break up an oil slick into tiny droplets. This allows the oil to diffuse in seawater, rather than remaining on the surface.

dispersion: A collective term for the spread of material in the biosphere by various physical processes such as wind, waves, currents, turbulence, and the like. Also refers to the process whereby water absorbs oil, breaking up an oil slick into tiny droplets. This can occur naturally or be hastened with the aid of chemical dispersants.

EEZ: Exclusive Economic Zone, as established by the 1982 Law of the Sea Convention (LOSC). An area of the sea beyond the territorial waters over which a state has sovereign rights for the purposes of exploration, exploitation, and resource management.

emulsification: The process by which wind buffets the surface of the sea producing foam and spray. This allows oil on the surface to absorb water. When enough water has been absorbed, the slick breaks up into brown masses called **mousse** or "chocolate mousse."

environmental echoes: Special public topics or conferences on environmental issues take place via FidoNet, an amateur computer network accessible via modem and personal computer. The Earth Echo is one example of such a network.

estuary: Area where fresh water at the mouth of a river mixes with the salt water of the sea. Estuaries are among the richest, most productive, and most intensively used habitats on the planet. The Chesapeake Bay is the largest estuary in the United States.

eutrophication: A process in which a large influx of nutrients (often from sewage) in a body of water causes abnormally rapid growth of algae; when the algae later dies and decomposes, oxygen in the water is depleted, which in turn kills marine animals such as fish that depend on dissolved oxygen.

evaporation: As regards oil in the marine environment, this term refers to the process whereby volatile hydrocarbons escape from an oil slick and diffuse into the air. Evaporation of oil on seawater is greatest during the first few hours after a spill. The oil slick gradually reduces to tar-like clumps as evaporation progresses. The rate of evaporation depends on many factors including ambient temperature, wave action, and the viscosity of the oil. Evaporated oil is an air pollutant that eventually may return to the sea in rainfall.

fallout: Particulate matter that gradually descends from the atmosphere after an explosion; often used to refer to radioactive particles from nuclear explosions.

GaiaNet: The term used to refer to the information network used by the worldwide environmental community. Can be accessed via FidoNet, BITNET, Usenet, or Internet by using a personal computer and a modem.

GESAMP: United Nations Joint Group of Experts on the Scientific Aspects of Marine Pollution. One of the key international scientific bodies in determining global marine environment policy.

gray: A measurement of the absorption rate of radiation by an organic substance, usually human or animal tissue.

halogenated hydrocarbons: Hydrocarbons to which halogens such as chlorine, bromine, or iodine have been added. Chlorinated hydrocarbons include polychlorinated biphenyls, known as PCBs, and the benzene-oxygen-chlorine combination known as dioxin. Halogenated hydrocarbons tend to accumulate in the environment, and most cause cancer, anemia, skin damage, and reproductive problems.

hydrocarbons: Compounds consisting of hydrogen and carbon. These are the building blocks of organic matter. Methane, benzene, ethylene, xylene, and toluene are all different types of hydrocarbons. Synthetic or processed hydrocarbons (such as **halogenated hydrocarbons** and petroleum products) are often toxic.

hypertrophication: *See* **eutrophication.**

ITOPF: International Tanker Owners Pollution Federation (*See also* **TOVALOP**).

joule: In the metric system, a unit of energy: the amount of energy required to move a kilogram mass one meter in one second.

LOSC: Law of the Sea Convention, a body of law that took a decade to create and promises to provide a global foundation for marine policy.

MARPOL: International Convention for the Prevention of Pollution from Ships, the international agreement that underlies national laws to prevent marine pollution in countries around the world.

mousse: Describes the brown, weathered mass formed by the action of wind and waves on an oil slick.

OILPOL: International Convention for the Prevention of Pollution of the Sea by Oil.

organochlorine: Organic chlorine compound.

outfall: Pipe that carries sewage into the sea.

pathogen: Microorganism capable of causing disease.

PCBs: Polychlorinated biphenyls; made by combining chlorine with phenol, a type of hydrocarbon. Extremely persistent halogenated hydrocarbon compounds that are chemically similar to DDT and its by-products. PCBs are a waste product of many industrial processes, especially the manufacture of paint, plastics, adhesives, and electrical equipment. Although great quantities of PCBs have been released into waterways or stored in canisters underground, 98 percent of the PCBs currently in the ocean entered the marine environment via the atmosphere.

persistence: In ecology, a term used to describe the resistance of certain substances—among them DDT, chlorinated and halogenated hydrocarbons—to chemical and microbial disintegration. These substances tend to remain in the food chain for long periods of time, and accumulate in the fatty tissues of animals.

persistent oil: Includes crude oil, heavy fuel oil, heavy diesel oil, and lubricating oil. This term is used in a number of pollution compensation agreements and conventions.

photodegradable: Capable of being degraded by exposure to light. In particular, photodegradable plastics break down when exposed to the ultraviolet rays of the sun. Such plastics are used for beverage six-pack rings, among other products.

phytoplankton: Free-floating microscopic plants. A single cubic yard of sea water can contain as many as 200,000 phytoplankton. Because they rely on photosynthesis, phytoplankton can only exist in the uppermost 330 feet of the sea.

plankton: A collective term for the microscopic plants (phytoplankton) and marine animals (zooplankton) that exist in huge quantities in the sea and which form the basis of the earth's complex food chain.

pollution: Direct or indirect introduction by man of substances or energy that result in deleterious effects on the environment, such as harm to living things.

polychlorinated biphenyl: *See* PCB.

pp: An abbreviation ("parts per") often used as a prefix in measurements of the concentration of toxins in seawater, for example **ppm** (parts per million), **ppb** (parts per billion), and **ppt** (parts per trillion).

primary treatment: Initial treatment of raw sewage which removes solids and allows any remaining particles to settle, producing a less offensive substance. Nutrients, viruses, and most toxic chemicals remain part of the effluent.

radiation: Energy released by radioactive decay (the disintegration of the nucleus of an unstable atom).

radionuclide: Radioactive isotope of an element; an unstable form of the element with a different number of neutrons in the nucleus than in the more common, stable form.

red tide: *See* **bloom.**

salinity: Proportion of salt in sea water.

secondary treatment: Sewage treatment process that breaks down the organic matter in sewage, thus reducing the need for oxygen for further decomposition if the sewage is dumped in the sea. The effluent resulting from secondary treatment still contains nutrients, soluble chemicals, and viruses.

sedimentation: The process whereby particles settle on the sea floor.

sievert (Sv): A unit of measurement of biological damage caused by different types of radiation. Equal to 1 **gray** for beta and gamma particles, 10 grays for neutrons, and 20 grays for alpha particles.

slick-licker: Boom or other mechanical device designed to contain oil slicks and then remove oil from the sea.

TBT: Tributyltin, also known as organotin. Used in barnacle-resistant paint for boat hulls. Toxins leach into coastal waters as boats treated with tributyltin remain moored for long periods of time, contaminating shellfish and other marine organisms. Now banned in many countries around the world.

TOVALOP: Tanker Owners Voluntary Agreement Concerning Liability for Oil Pollution. This is a worldwide agreement to meet damage and cleanup costs in oil tanker accidents. It is administered by the International Tanker Owners Pollution Federation (ITOPF).

weathering: The process whereby an oil slick turns into an oil-in-water emulsion via the action of wind and wave, forming tar clots and **mousse.**

zooplankton: Just above phytoplankton in the food chain, zooplankton range in size from single-celled organisms to jellyfish. They either graze on phytoplankton or are carnivorous, preying on other zooplankton.

Index

Acid rain, 46
ACOPS (Advisory Committee on the Protection of the Sea), 149, 186
ACOPS Year Book 1986–1987, 186
Advisory Committee on the Protection of the Sea (ACOPS), 149, 186
Aegean Captain, 42
Aegean Sea, 50
Aguilar, Patricio, 187
Alaska, 48
Alaska: Crude Awakening (video), 211
Alaska Conservation Foundation (ACF), 149–150
The Almanac of Renewable Energy, 61
The Almanac of Science and Technology: What's New and What's Known, 61
Alpha particles, 23
American Cetacean Society (ACS), 150
American Oceans Campaign (AOC), 58–59, 150–151
America's Biggest Oil Spill (video), 211–212
America's Trash Hits Home (video), 212
Amoco Cadiz, 41, 46
Anaerobic decomposition, 6
And Two If by Sea: Fighting the Attack on America's Coasts: A Citizen's Guide to the Coastal Zone Management Act and Other Coastal Laws, 201
Anti-nuclear activists, 55–56
Aquatic Habitat Institute, 151
Are You Swimming in a Sewer? (video), 212
Argo Merchant, 40
Ariadne, 45
Arthur J. Morris Law Library, University of Virginia, 183–184
Assimilative capacity, ocean's, 28
Assimilative Capacity of U.S. Coastal Waters for Pollutants, 61
Atlantic Empress, 42
Atmosphere, 19
Atomic bombs, 33
Atomic Energy Commission, 33

Backer, Terry, 53–54
Baines, John D., 195
Baltic Marine Environment Protection Commission–Helsinki Commission, 151
Barnett, Judith B., 186
Bascom, Willard, 133–134, 204
Bay Conservation and Development Commission (BCDC), 152
Bay Information Network (BIN), 151
Bay Institute of San Francisco, 152
Baykeeper, 152
Baymen, Our Waters Are Dying (film), 213
Beach Clean-Up, 44
Beach Cleanup Campaign, 64
Becquerels, 27
Before It's Too Late: A Scientist's Case FOR Nuclear Energy, 209–210
Beta particles, 23
Betelgeuse, 42
Between the Devil and the Deep Blue Sea (video), 213
The Big Spill (video), 213–214
Bikini Atoll, 33, 38
Billecci, Tom, 54
Bioaccumulation, 13, 29

Biodegradable waste, 5
Biological oxygen demand (BOD), 6
Bioremediation, 22
Bioremediation for Marine Oil Spills, 191
Black Water Time Bomb (video), 214
Blockstein, David E., 205
Bloom, 6
Blow Out: A Case Study of the Santa Barbara Oil Spill, 209
Blowouts, 19
Booker, F., 208
Booms and mechanical devices, 21
Braer, 50
Bright, Michael, 195
British National Committee on Oceanic Research, Marine Pollution Subcommittee, 196
British Nuclear Fuels (BNFL), 44
Brown, Lester, 186
Brownlee, Shannon, 196
Bruntland, Gro Harlem, 54–55
BT Nautilus, 47
Bulloch, David K., 196
Burning oil pollution, 21

Cadmium, 17
Cahan, V., 207
Caldicott, Helen, 55–56
California Coastal Commission, 153, 186
California Coastal Conservancy, 153
California State Water Resources Control Board, 153–154
Canada–United States Environmental Council (CUSEC). *See* Defenders of Wildlife
Capper, John, 196
Carson, Rachel, 29, 56
Catena, John G., 203
Center for Marine Conservation, 44, 134, 154, 186
Center for Oceans Law and Policy, 154–155
Center for Short-Lived Phenomena (CSLP), 155
Centre for Our Common Future, 55
Chabreck, Robert H., 196
Challenge of the Seven Seas, 65
Champ, Michael A., 187
Chasis, Sarah, 56
Chemical dispersants, 21
Chesapeake: Living Off the Land (video/film), 214
Chesapeake Bay, 214
Chesapeake Bay CleanUp Agreement, 43
Chesapeake Bay Foundation (CBF), 155
Chesapeake Waters: Pollution, Public Health and Public Opinion, 196
Chevron, 37
Chief Seattle, 31
Childers, Roberta, 190
Chlorinated hydrocarbons, 13
Christol, Carl Q., 187
Christos Bitas, 41
A Citizen's Guide to Plastics in the Ocean: More Than A Litter Problem, 65, 206
Clark, Robert B., 131, 197
Clean Seas, 156
Clean-up
  heavy metals, 17–18
  marine debris, 10
  oil pollution, 21–22
  radioactive waste, 27
  sewage pollution, 7
  toxic chemicals, 14
Clean Water Act, 45
Clean Waters Restoration Act of 1966, 36
Cleaning North America's Beaches: 1991 National Beach Cleanup Report, 205–206
Clinton, Bill, 50
The Closing Circle, 57
Coast Alliance, 156
Coastal Clean-Up (slides), 214–215
Coastal Connection, 190
Coastal Marshes: Ecology and Wildlife Management, 196–197
Coastal Pollution Hearing, 137
Coastal States Organization, 156–157
Coastal Zone Management Act (CZMA), 39, 47, 142
Coastal Zone Space, 61
COASTNET, 227
COASTWEEKS, 9–10
Cohen, Bernard L., 209

Commercialization of the Oceans, 202
Commission on Marine Science, 142, 192
Commoner, Barry, 57
CompuServe, 227
Computer networks and databases, 227–230
CONCAWE. *See* Oil Companies' European Organization for Environmental and Health Protection (CONCAWE)
Conference on the Status of Knowledge, Critical Research Needs and Potential Research Facilities . . . , 36
"Congress Tackles Ocean Plastic Pollution," 205
Connecticut Fishermen's Association, 53–54
Construction rubbish dumping, 31
The Control of Certain Pollutants in Navigable Waters, 95–96
The Control of Certain Pollutants in New York Harbor, 95
Convention on the Liability of Operators of Nuclear Ships, 35
Convention on the Prevention of Marine Pollution by Dumping of Wastes and Other Matter, 74–78
Copper, 17
Copper sulfate dumping, 35
Corson, Walter H., 187
Cosmos satellite, 39
Cottingham, David, 192
Council on Ocean Law, 157
Couper, Alastair, 187
Cousteau, Jacques-Yves, 57–58, 131–132, 143–144, 197
Cousteau, Jean-Michel, 58
Cousteau Collection (video), 215
Cousteau Odyssey (video), 215–216
Cousteau Society, Inc., 157–158
Cover Up: What You Are Not Supposed To Know about Nuclear Power, 210
CRISTAL (Contract Regarding an Interim Supplement to Tanker Liability for Oil Pollution), 42
The Cry of the Beluga (video), 216
Curie, 27
Dana Optima, 44
Danger at the Beach (video), 216
Danson, Ted, 58–59, 134
Davidson, Art, 207
DDT: Scientists, Citizens, and Public Policy, 207
DDT (dichlorodiphenyltrichloroethane), 11, 12, 13, 14, 31, 32
Deep Ocean Technology, 60
D'Elia, Christopher F., 197
Denis, Corinne, 197
Dialog Information Services, 227
Dickson, David, 206
Dioxins, 11, 12, 13
Directory of Environmental Organizations, 227–228
Directory of NOAA Library and Information Network, 188
The Disposal of Industrial and Domestic Wastes, 61
"The Disposal of Waste in the Ocean," 204
Do We Really Want To Live This Way? (video), 217
Dold, Catherine, 207
Don't Mess With Texas Beaches (video), 216–217
Don't Teach Your Trash to Swim (video), 217
Douglas, Marjory Stoneman, 59–60
Down the Shore (video), 217
"The Dragon Lady," 204
Drifnets, 64, 99, 101–102
The Drowning Bay (video), 217–218
Dunlap, Thomas R., 207
The Dying Sea, 195–196

E-Town, 231
Earle, Sylvia, 60
Earth, 2–3
Earth Island Institute, 158
Earth Summit, United Nations (UNSET), 49
Earthwatch, 158
Ebb Tide for Pollution: Actions for Cleaning Up Coastal Waters, 56, 198
"Ecocatastrophe!," 198
EcoNet, 228
EcoNews, 230

EDF (Environmental Defense Fund), 46, 159–160
The Effects of Atomic Radiation on Oceanography and Fisheries, 193–194
Effects of Man's Activities on the Marine Environment, 193
The Effects of Marine Pollution: Some Research Needs, 196
Effluent, 5, 32
Ehrlich, Paul, 143, 198
Eichbaum, William, 200
Eleni V, 41
Encyclopedia of Environmental Science and Engineering, 189
Enskeri, 40
Entanglement Network, 64
Entanglement Network Coalition, 159
Enviroline, 228
Environmental Abstracts, 187
Environmental Action, Inc., 159
Environmental Bibliography, 228–229
Environmental crisis, 107
Environmental Defense Fund (EDF), 46, 159–160
Environmental Directions, 230
Environmental ethics, 29
Environmental Hazards: Water Pollution, a Bibliography, 188
Eros 2000 program, 45
Esso Bernicia, 42
Eutrophication, 6
The Everglades: River of Grass, 59
Everglades National Park, 59
Exxon Valdez, 21, 22, 47, 48, 49, 51, 56, 139, 211–212, 213, 219

Federal Council for Science and Technology, 192
Federal Plan for Ocean Pollution Research, Development and Monitoring, 194
Federal Water Pollution Control Act Amendments, 39
Fine, John C., 198
First International Conference on Waste Disposal in the Marine Environment, 34, 132
First North Sea Conference, 45
500 Million Years Beneath the Sea (video/film), 218
Florida Everglades, 49
The Frail Ocean: A Blueprint for Change in the 1990s and Beyond, 201
Friedman, Wolfgang, 198
Friends of the Earth, 160
The Friends of the Everglades, 59–60, 160–161
From Sea to Shining Sea (video), 218–219
Furness, Robert W., 207
The Future of the Oceans, 198

Gamma rays, 23
Geneva Conference on the Law of the Sea, 34
Gerlach, Sebastian A., 137, 185, 198
GESAMP. *See* Joint Group of Experts on the Scientific Aspects of Marine Pollution (GESAMP)
Get Oil Out, Inc. (GOO), 161
Getting to the Bottom of It: Threats to Human Health and Environment from Contaminated Underwater Sediments, 199
Gill, C., 208
Gist, Ginger L., 208
The Global Ecology Handbook: What You Can Do about the Environmental Crisis, 187
Global Marine Pollution Bibliography: Ocean Dumping of Municipal and Industrial Wastes, 187
Goldberg, Edward D., 60–61, 136–137, 199
Goldsmith, Edward, 199
Golob, Richard S., 61–62
Golob's Oil Pollution Bulletin, 21, 61, 190
Goodavage, Maria, 199
Gore, Al, 50
Gourlay, K. A., 200
Government publications, 191–195
Grays, 27
Greenpeace USA, 43, 161–162
Grossman, Karl, 210
A Guide to Marine Pollution, 61
Gulf of Mexico, 33, 37

Half-life, 23
Handbook on Ocean Politics and Law, 203
Hanson, M. Bradley, 187
Hazardous Materials Intelligence Report, 61, 190
Heal the Bay, 162
The Health of the Oceans, 61, 199
Heavy metals, 119, 137–138, 207
  cleaning up, 17–18
  effects on living organisms, 16–17
  entering ocean, 15–16
  monitoring, 17
  solutions, 18
Heavy Metals in the Marine Environment, 207
Hellenic Marine Environment Protection Association (HELMEPA), 162–163
Herz, Michael Joseph, 62
Heyerdahl, Thor, 34, 39, 62–63, 130, 144, 200
Hickel, Walter, 142
High-level radioactive wastes, 25
Hood, Donald W., 200
Horton, Tom, 200
How To Kill an Ocean, 63, 200
Hoyle, Russ, 204

Illegal dumping, 54
Imperiled Planet: Restoring Our Endangered Ecosystems, 199
Impingement of Man on the Oceans, 200
In the Wake of the Exxon Valdez: *The Devastating Impact of the Alaska Oil Spill,* 207–208
Incredible Facts about the Ocean, Vol. 3: How We Use It, How We Abuse It, 202
Inform, 163
Interagency Committee on Ocean Pollution Research, 192
Intergovernmental Maritime Consultative Organization (IMCO), 33
International Agreements on Conservation of Marine Resources, with Special Reference to the North Pacific, 203
International Atomic Energy Agency, 26
International Atomic Energy Commission, 39, 40
International Center for the Solution of Environmental Problems, 163
International Commission on Radiological Protection (ICRP), 140–141
International Commission on Radiological Units and Measurements (ICRU), 26
International Congress of Nuclear Scientists, 35
International Convention for the Prevention of Pollution from Ships (MARPOL), 39, 46, 63, 78–84
International Convention for the Prevention of Pollution of the Sea by Oil (OILPOL), 34, 71–74
International Convention on Civil Liability for Oil Pollution Damage (CLC), 37
International Convention on the Establishment of an International Fund for Oil Pollution Damage, 38
International Convention Relating to Intervention on the High Seas in Cases of Oil Pollution Casualties, 37
International conventions and agreements, 68–92
International Environmental Reporter, 188
International Handbook of Pollution Control, 188
International Marinelife Alliance, 163–164
International Maritime Conference, 32
International Maritime Organization (IMO), 47, 164
International Meeting on Wildlife and Oil Pollution, 41
International Ocean Disposal Symposium. *See* International Ocean Pollution Symposium

International Ocean Pollution Symposium, 165
The International Oceanographic Foundation (IOF), 164–165
International Organizations and the Law of the Sea Documentary Yearbook, 189
International Tanker Owners Pollution Federation Ltd. (ITOPF), 165–166, 208
Iraq, 48

Johnston, R., 200
Joint Conference on Prevention and Control of Oil Spills, 138–139
Joint Group of Experts on the Scientific Aspects of Marine Pollution (GESAMP), 4, 130, 146, 166, 192

Keeble, John, 208
Keep Britain Tidy Group. *See* Tidy Britain Group
Kennedy, John F., 129
Kester, Dana R., 204
Ketchum, Bostwick H., 63, 204
Kormondy, Edward J., 188

Land-based radioactivity, 24
Law of the Sea, 68
Law of the Sea: Protection and Preservation of the Marine Environment, 194
Law of the Sea Convention (LOSC), 39, 43, 65, 84–90
Laws of Ecology, 57
Lead, 16–17
Legal framework, 68–106
Libraries, 183–184
Life in the Oceans, 204
Liptak, Karen, 201
The Living Ocean: Understanding and Protecting Marine Diversity, 203
Living organisms, effects on
  heavy metals, 16–17
  oil, 20
  radiation, 25–26
  toxic chemicals, 13–14
  marine debris, 9
Living with the Spill (video), 219
Los Angeles County Sanitation District, 32
LOSC (Law of the Sea Convention), 39, 43, 65, 84–90
Louisiana Universities Marine Consortium, 166–167
Low-level radioactive wastes, 25
Lynch, Kevin, 205

Managing Troubled Waters: The Role of Marine Environmental Monitoring, 193
Manville, Albert, II, 63–64
Marine and Coastal Education Resources Directory, 186
Marine Conservation News, 191
Marine contamination, 4
Marine debris, 7–11, 113–114, 205–206
  cleaning up, 10
  effects on living organisms, 9
  and entanglement, 219
  entering ocean, 8
  monitoring, 9–10
  points of view, 134–135
  solutions, 10–11
  types of, 7–8
Marine Debris and Entanglement (slides), 219
Marine Debris Bibliography, 187–188
Marine Debris Educational Materials, 186
Marine Debris Information Office (MDIO). *See* Center for Marine Conservation
Marine Forum for Environmental Issues, 167
Marine incineration, 12
Marine Plastic Pollution Control Act (MPPCA), 99, 100
Marine pollution, 4
  changing perception, 28–29, 141–147
  defined, 3–4, 129–132
  in Europe, 141
  heavy metals, 15–18
  marine debris, 7–11
  oil, 18–23
  points of view, 141–147
  radioactive materials, 23–28

sewage, 5–7
toxic chemicals, 11–15
Marine Pollution, 197, 200
Marine Pollution: Diagnosis and Therapy, 185, 198–199
Marine Pollution Bulletin, 191
Marine Pollution Control Unit (of the DTp) (MPCU), 167–168
Marine Pollution Monitoring: Strategies for a National Program, 61
Marine Protection, Research, and Sanctuaries Act of 1972, 39, 40
Marine Science Institute, 168
Marine Science Journals and Serials: An Analytical Guide, 186
Marine Wildlife Entanglement in North America, 206
MARPOL (International Convention for the Prevention of Pollution from Ships), 39, 46, 63, 78–84
Marshall, Eliot, 205
Marx, Wesley, 201
Mercury, 17
Mercury poisoning, 1–2, 137–138
Methods of Detection, Measurement and Monitoring of Pollutants in the Marine Environment, 61
Methy ethylketone (MEK) leak, 44
Mielke, James E., 193
Millemann, Beth, 201
Miller, E. Willard, 188
Miller, Ruby M., 188
Minamata disease, 1–2, 34
Monitoring
marine debris, 9–10
oil pollution, 20–21
radiation, 26–27
sewage, 6–7
Mont Louis, 44
Moorcraft, Colin, 201
"Murky Waters," 199–200
Mussel Watch, 61
Must the Seas Die?, 201
"Mystery Disease Strikes Europe's Seals," 206–207

National Academy of Sciences, 35, 193–194
National Advisory Council on Oceans and Atmosphere, 194
National Audubon Society, 168
National Coalition Against the Misuse of Pesticides, 169
National Coalition for Marine Conservation, 169
National Coastal Caucus, 150
National Environmental Policy Act (NEPA), 37
National Marine Pollution Plan, 1981–1985, 192
National Multi-Agency Oil and Hazardous Materials Contingency Plan, 36
National Ocean Goals and Objectives for the 1980s, 194
National Ocean Industries Association, 169
National Oceanic Society, 62
National Oceanographic and Atmospheric Administration Library, 184
National Oceanographic and Atmospheric Administration (NOAA), 51, 61, 170, 188, 194
National Research Council, 192, 194, 195
National Resources Defense Council (NRDC), 51
National Science Foundation (NSF), 170
National Sea Grant Depository, 184
National Toxics Campaign (NTC), 170–171
Natural Resources Defense Council (NRDC), 56, 171
Nature Conservancy Council, 41, 208
Nelson-Smith, A., 208
Neptune's Revenge: The Ocean of Tomorrow, 202
Netherlands Institute for the Law of the Sea (NILOS), 171–172, 189
"The New Dead Sea: Persian Gulf Oil Spill Advances Ecological Clock," 208
"The New Guardians," 207
New York Harbor, sewage in, 32
1991 Beach Pollution Report, 205
NOAA. *See* National Oceanographic and Atmospheric Administration
North Sea Directorate, 172
NRDC Coastal Project, 56

NRDC (National Resources Defense Council), 56, 171
Nuclear chain reaction, 33
Nuclear Information and Resource Service (NIRS), 172–173
Nuclear Madness: What You Can Do, 55
Nuclear-powered vehicles and devices, 25
The Nuclear Question, 210
"Nutrient Enrichment of the Chesapeake Bay," 197

OCA/PAC. *See* United Nations Environment Programme (UNEP)
Ocean characteristics, 2–3
The Ocean Dumping Act of 1972, 39, 96–99
Ocean Dumping Convention, 74–78
Ocean Dumping of Industrial Wastes, 204–205
The Ocean Planet: The Death of the Mississippi (video), 219–220
Ocean Politics and Law: An Annotated Bibliography, 189
Ocean pollution, 107
Ocean Pollution: What You Should Know (slides), 220
Ocean Pollution in the Marine Environment: A Legal Bibliography, 187
Ocean Pollution Research Center, 173
The Ocean Sink (video), 220
Ocean Transport of Radioactive Materials conference, 35
The Ocean World of Jacques Cousteau, 197
Oceanic Abstracts, 229
The Oceanic Society. *See* Friends of the Earth
Oceanography 1960–1970, 193
Oceanography—The Ten Years Ahead, 192
The Oceans, Our Last Resource, 201
Oceans in Peril, 198
Oceanus, 191
Offshore oil drilling, 31, 33
Offshore oil platform, 19
O'Hara, Kathryn Jean, 64–65, 206
Oil Companies' European Organization for Environmental and Health Protection (CONCAWE), 173–174
Oil Companies International Marine Forum (OCIMF), 174
Oil drilling, offshore, 31, 33
Oil in the Sea: Inputs, Fates and Effects, 195
Oil pollution, 18–23, 119–122, 207–209
  cleaning up, 21–22
  effects on living organisms, 20
  entering ocean, 19
  monitoring, 20–21
  points of view, 138–139
  solutions, 22–23
Oil Pollution Act of 1961, 35
Oil Pollution Act of 1990, 47, 102–106
Oil Pollution and Marine Ecology, 208
Oil Pollution Manual, 208
Oil Spill Intelligence Report, 229
Oil Spills: A Coastal Resident's Handbook, 209
Oil Spills: Just a Cost of Doing Business, 209
Oil tankers, 22
OILPOL (International Convention for the Prevention of Pollution of the Sea by Oil), 34, 71–74
Omohundro, John T., 209
Our Common Future, 55, 204
Our Nation and the Sea, 192
"Our Threatened Planet," 202
Out of the Channel: The Exxon Valdez *Oil Spill in Prince William Sound,* 208
Outer Continental Shelf Lands Act, 34

Parc Oceanique Cousteau, 58
Paris Commission, 174–175
Paris Convention, 40
Park, P. Kilho, 187, 204
Passion & Commitment (audio cassette), 220–221
PCBs (polychlorinated biphenyls), 11, 12, 36
Pearce, Fred, 209
Pell, Claiborne, 36, 65

Persistent Marine Debris: Challenge and Response: The Federal Perspective, 192, 206
Petroleum, 18. *See also* headings beginning with Oil
Petroleum products, reducing demand for, 22–23
Pfafflin, J. R., 189
Phytoplankton overgrowth, 6
Plastic debris, 8
"Plastic Reaps a Grim Harvest in the Oceans of the World," 206
Plastics development, 33
Poisoners of the Seas, 200
Pollution Abstracts, 229
Polyaromatic hydrocarbons (PAHs), 20
Population statistics, 106, 107
Ports and Waterways Safety Act, 39
Power, Garrett, 196
President's Panel on Oil Spills, 36
The Protected Ocean: How To Keep the Seas Alive, 201
Protecting Coastal Waters: A Community Approach (video), 221
Protecting the Oceans, 195
Public Law 100-220, 99–102
Public Law 101-380, 102–106

Radio Earth Island, 231
Radio series, 230–231
Radioactive fallout, 24
Radioactive materials, 23–28, 40, 122–128, 209–210
  cleaning up, 27
  disposing of, 25
  effect on living organisms, 25–26
  entering ocean, 24–25
  measures of, 27
  monitoring, 26–27
  points of view, 140–141
  solutions, 27–28
  therapeutic use, 26
Radioactive Waste Disposal into Atlantic and Gulf Coastal Waters, 35, 194
Radioactivity in the Marine Environment, 193, 194
Radioactivity sickness, 26
Radionuclides, 33
Radiosotopes, 24
Rads, 27
Radwaste, 210
Rainbow, Philip S., 207
Raum, A. M., 132–133
Red tide, 6, 99, 102
Reducing World Pollution: A Compilation of United Nations and U.S. Government Documentary Materials, 189
Reef Relief, 175
Regional Seas Activity Center and Regional Seas Research Studies. *See* United Nations Environment Programme (UNEP)
Regulations for the Safe Transport of Radioactive Materials, 39
Remnants of Eden (video), 221
Rems, 27
Report of the Swedish Royal Commission on Natural Resources, 142
Resources for the Future, Inc., 175
Response to Marine Oil Spills, 208
Restless Oceans, 203–204
Revelle, Roger, 132
Review of the State of the Marine Environment, 146
Rivers, oil pollution, 19
The Rivers and Harbors Act, 31, 92
Robinson, W. Wright, 202
The Role of the Ocean in a Waste Management Strategy: A Special Report to the President and Congress, 195
Rome Conference on Oil Pollution of the Sea, 141

Save the Bay, 176
Saving Our Wetlands and Their Wildlife, 201
Science and Survival, 57
Scientific Problems Relating to Ocean Pollution, 61
Scripps Institution of Oceanography, 176
The Sea: Mysteries of the Deep (video), 221–222
The Sea around Us, 56
Sea Grant College program, 36, 65, 99, 101, 176–177
"The Sea Has Its Limits," 197
The Sea in Danger, 197
Sea of Marmara, 31

Seabrook, New Hampshire nuclear power plant, 49
SEAPRESS, 229
Seas and Oceans—An Evening of Choices for the Future Symposium (video), 222
Seas At Risk, 177
Seas Under Siege (video), 222–223
Second North Sea Conference, 45
Sedge, Michael H., 202
Sentinel of the Sea (video), 223
Seredich, John, 189
Sewage, 204–205
  cleaning up, 7
  in marine waters, 108–110
  monitoring, 6–7
  points of view, 132–134
  treatment, 5–6, 111–113
Sewers, 31, 212–213
Shapiro, Fred C., 210
Shetland Islands, 50
Shinoussa, 48
Ships, as source of marine pollution, 19, 107–108
Shivers, Frank R., Jr., 196
Sibthorp, M. M., 142
Sierra Club Clean Coastal Waters Task Force, 177–178
Sierra Club Legal Defense Fund, 44, 48, 54
Sieverts, 27
The Silent World, 58
Simon, Anne W., 202
Sixteenth Governing Conference, 38
Skidaway Institute of Oceanography, 178
Silent Spring, 56
Sludge, 5
"The Sludge Factor," 205
Smithsonian Institution, 178
SNAP-9A satellite, 35
Soafer, Abraham, 43
Soper, T., 208
Spencer, Page, 209
The State of the Marine Environment, 192
State of the World: A Worldwatch Institute Report on Progress toward a Sustainable Society, 186
Steinhart, Carol E., 209
Steinhart, John S., 209
Stella Maris, 38
Stone, Roger D., 202
"Stopping Coastline Pollution at the Sewer and at the Farm," 196
Student Environmental Action Coalition, 178–179
Study of Critical Environmental Problems (SCEP), 38
Summerland Field, 31
Sustainable development, 55
Swann Chemical Company, 32

Technical Conference on Marine Pollution and Its Effects on Living Resources and Fishing, 38
Television series, 230–231
Tesar, Jenny, 203
Testing the Waters: A National Perspective on Beach Closings, 56, 205
Testing the Waters III: Closings, Costs and Cleanup at U.S. Beaches, 51
Texas Shores—Saving What's Left (video), 223
Third Session of the Advisory Committee on Marine Resources Research, 133
Thorne-Miller, Boyce, 203
Thornton, James, 43
Threatened Oceans, 203
Tide, 6
Tide, 191
Tidy Britain Group, 179
The Times Atlas and Encyclopedia of the Sea, 187
"To Stop Spills, Punishment Must Cost More than Prevention," 207
Tomasevich, J., 203
Torrey Canyon, 4, 21, 36, 37
TOVALOP (Tankers Owners Voluntary Agreement concerning Liability for Oil Pollution), 37. *See also* ITOPF, International Tanker Owners Pollution Federation Ltd.
Toxic chemicals, 114–119, 206–207, 216
  cleaning up, 14
  effects on living organisms, 13–14
  entering ocean, 12–13

points of view, 135–137
solutions, 14–15
Toxic Substances Control Act, 40
Trashing the Oceans & Port of Newport (Oregon) (video), 223–224
Treasures of Neptune: Klondike on the Ocean Floor (video), 224
Treatment, sewage, 5–6, 111–113
Treaty Banning Nuclear Weapons Testing, 35
Tributyltin (TBT), 43, 46
Troubled Waters (video), 224
Trzyna, Thaddeus C., 190
Turning the Tide, 200
Turning the Tide: Keeping Pollution at Bay (video), 224–225

The Undersea World of Jacques Cousteau, 58
Underwater Marine Debris Collection (slides), 225
UNESCO, 129
Union of Concerned Scientists, 50, 147
Union Oil. *See* Unocal
United Nations, 194
United Nations Conference on the Human Environment (UNCHE), 68, 90–92
United Nations Convention on the Law of the Sea (LOSC). *See* Law of the Sea Convention (LOSC)
United Nations Earth Summit (UNSET), 49
United Nations Educational, Scientific, and Cultural Organization (UNESCO), 129, 180
United Nations Environment Programme (UNEP), 179–180
United Nations Food and Agriculture Organization (FAO), 38, 61
United Nation's General Assembly, 54
United States Coast Guard, 180
United States Department of Commerce, 195
United States Department of Energy, 180–181
United States Environmental Protection Agency (US EPA), 25, 181
United States Geological Survey, 181
United States–Japan Fishery Agreement Approval Act of 1987, 99–100
United States National Advisory Committee on Oceans and Atmosphere, 195
Unocal, 36–37, 44, 48, 54
U.S. EPA, 25, 181
U.S. Fish and Wildlife Service, 135–136
U.S. marine pollution laws, 92–106
U.S. Ocean Policy in the 1970s: Status and Issues, 195

The Voyage of the Sanderling, 202

Wang, James C., 189, 203
Ward, Barbara, 145
Warren Spring Laboratory (WSL), 182
Waste Not, Want Not (video), 225
The Wasted Ocean, 196
Wasting Away, 205
Water Quality Improvement Act, 37
Water volume, ocean, 2
Weiss, Ann E., 210
Weisskopf, Michael, 206
The WELL, 230
Whales, 150, 216
What Price Progress? (film), 225–226
Where Have All the Dolphins Gone? (video), 226
Whipple, A. B. C., 203
White Silk & Black Tar: A Journal of the Alaska Oil Spill, 209
Whose Sea Is This? (video), 226
The Wilderness Society, 182
"Wildlife Choked by World's Worst Oil Slick," 209
Windscale nuclear power plant, 34
Woods Hole Oceanographic Institution, 182–183
Workshop on the Fate and Impact of Marine Debris, 44
World Book Yearbook, 129

World Commission on Environment and Development, 204
World Directory of Environmental Organizations: A Handbook of National and International Organizations and Programs, 190
World Environment Center, 183
The Wreck of the Torrey Canyon, 208
Wu, Norbert, 204

Year of the Ocean, 45
Your Resource Guide to Environmental Organizations, 189

Ziegler, E.N., 189

DATE DUE

APR 06 1996	
APR 29 1996	
NOV 29 1997	

GAYLORD PRINTED IN U.S.A.